BRITISH RAILWAYS POCKET BOOK No.4

ELECTRIC MULTIPLE UNITS

TI

The con

Multiple U ... on the

Railtrack ... urotunnel networks

Neil Webster

ISBN 1 902336 12 7

CONTENTS

Updates 2
Readers Comments 2
Organisation & Operation of Britain's Railway System 3
Using This Book 6
General Information 9
Electric Multiple Units
1. 25 kV a.c. 50 Hz overhead EMUS 13
2. 750 V d.c. third rail EMUS 54
3. Eurostar EMUS. 85
4. Service EMUS. 97
5. Vehicles Awaiting Disposal. 103
Livery Codes 104
Owner Codes 105
Operation Codes 106
Depot Type Abbreviations 107
Depot & Location Codes 107

UPDATES

An update to all the books in the *British Railways Pocket Book* series is published every month in the Platform 5 magazine, *Today's Railways*, which also contains news and rolling stock information on the railways of Britain, Ireland and Continental Europe. Rolling stock updates will also be found in other magazines specialising mainly in British matters, such as "Rail Express". For further details of *Today's Railways,* please see the advertisement inside the front cover of this book.

Information in this edition is intended to illustrate the actual situation on Britain's railways, rather than necessarily agree with TOPS and computer other records. Information is updated to 15 November 1999.

READERS' COMMENTS

With such a wealth of information as contained in this book, it is inevitable a few inaccuracies may be found. The author will be pleased to receive notification from readers of any such inaccuracies, and also notification of any additional information to supplement our records and thus enhance future editions. Please address comments to Neil Webster either at the address shown on the title page, or alternatively by fax (0114 255 2471) or e-mail (neil@platfive.freeserve.co.uk).

Please note: Both the author and the staff of Platform 5 regret they are unable to answer specific queries – either in writing or by telephone – regarding locomotives and rolling stock other than through the 'Q & A' section of *Today's Railways* magazine.

ORGANISATION & OPERATION OF BRITAIN'S RAILWAY SYSTEM

INFRASTRUCTURE & OPERATION

Britain's national railway infrastructure i.e. the track, signalling, stations and associated power supply equipment is owned by a public company – Railtrack PLC. Many stations and maintenance depots are leased to and operated by Train Operating Companies (TOCs), but some larger stations remain under Railtrack control. The only exception is the infrastructure on the Isle of Wight, which is nationally owned and is leased to the Island Line franchisee.

Trains are operated by TOCs over the Railtrack network, regulated by access agreements between the parties involved. In general, TOCs are responsible for the provision and maintenance of the locomotives, rolling stock and staff necessary for the direct operation of services, whilst Railtrack is responsible for the provision and maintenance of the infrastructure and also for staff needed to regulate the operation of services.

DOMESTIC PASSENGER TRAIN OPERATORS

The large majority of passenger trains are operated by the TOCs on fixed term franchises. Franchise expiry dates are shown in parentheses in the list of franchisees below:

Franchise	*Franchisee*	*Trading Name*
Anglia Railways	GB Railways Ltd. (until 4 April 2004)	Anglia Railways
Cardiff Railway	Prism Rail PLC (until 12 April 2004)	Cardiff Railways
Central Trains	National Express Group PLC (until 1 April 2004)	Central Trains
Chiltern Railways	M40 Trains Ltd. (until 20 July 2003)	Chiltern Railways
Cross Country	Virgin Rail Group Ltd. (until 4 January 2012)	Virgin Trains
Gatwick Express	National Express Group PLC (until 27 April 2011)	Gatwick Express
Great Eastern Railway	First Group PLC (until 4 April 2004)	First Great Eastern
Great Western Trains	First Group PLC (until 3 February 2006)	First Great Western
Inter City East Coast	GNER Holdings Ltd. (until 4 April 2004)	Great North Eastern Railway
Inter City West Coast	Virgin Rail Group Ltd. (until 8 March 2012)	Virgin Trains
Island Line	Stagecoach Holdings PLC (until 12 October 2001)	Island Line

LTS Rail	Prism Rail PLC (until 25 May 2011)	LTS Rail
Merseyrail Electrics	MTL Rail Ltd. (until 18 March 2004)	Merseyrail Electrics
Midland Main Line	National Express Group PLC (until 27 April 2006)	Midland Mainline
Network South Central	Connex Rail Ltd. (until 25 May 2003)	Connex South Central
North London Railways	National Express Group PLC (until 1 September 2004)	Silverlink Train Services
North West Regional Railways	First Group PLC (until 1 April 2004)	First North Western
Regional Railways North East	MTL Rail Ltd. (until 1 April 2004)	Northern Spirit
Scot Rail	National Express Group PLC (until 30 March 2004)	ScotRail
South Eastern	Connex Rail Ltd. (until 12 October 2011)	Connex South Eastern
South Wales & West	Prism Rail PLC (until 12 April 2004)	Wales & West Passenger Trains
South West	Stagecoach Holdings PLC (until 3 February 2003)	South West Trains
Thames	Victory Railways Holdings Ltd. (until 12 April 2004)	Thames Trains
Thameslink	GOVIA Ltd. (until 1 April 2004)	Thameslink Rail
West Anglia Great Northern	Prism Rail PLC (until 4 April 2004)	WAGN

The above companies may also operate other services under 'Open Access' arrangements.

The following operators run non-franchised services only:

Operator	*Trading Name*	*Route*
British Airports Authority	Heathrow Express	London Paddington–Heathrow Airport
West Coast Railway Co.	West Coast Railway	Fort William–Mallaig

INTERNATIONAL PASSENGER OPERATIONS

Eurostar (UK) operates international passenger-only services between the United Kingdom and continental Europe, jointly with French National Railways (SNCF) and Belgian National Railways (SNCB/NMBS). Eurostar (UK) is a subsidiary of London & Continental Railways, which is jointly owned by National Express Group PLC and the British Airports Authority.

In addition, a service for the conveyance of accompanied road vehicles through the Channel Tunnel is provided by the tunnel operating company, Eurotunnel, which is a partnership between The Channel Tunnel Group Limited and France-Manche SA.

FREIGHT TRAIN OPERATIONS

Freight train services are operated under 'Open Access' arrangements by the following companies:

English Welsh & Scottish Railway (EWS)
Freightliner
Direct Rail Services (a subsidiary of British Nuclear Fuels Ltd.)
Mendip Rail

USING THIS BOOK

LAYOUT OF INFORMATION

25 kV a.c. 50 Hz overhead Electric Multiple Units (EMUs) are listed in numerical order of class number, then in numerical order of set number – using official numbers as allocated by the Rolling Stock Library (RSL – the national rail vehicle registry). Individual 'loose' vehicles are listed in numerical order after vehicles formed into fixed formations. Where numbers carried are different to those officially allocated, these are noted in class headings where appropriate.

750 V d.c. third rail EMUs are listed in numerical order of set numbers actually carried. Some of these differ from the six digit numbers allocated by the RSL.

Where sets or vehicles have been renumbered since the previous edition of this book, former numbering detail is shown alongside current detail.

Each entry is laid out as in the following example:

Set No.	*Detail*	*Livery*	*Owner*	*Operation*	*Depot*	*Formation*
1706	†	**CX**	A	*SC*	BI	76094 63035 70713 76040

CLASS HEADINGS

Principal details and dimensions are quoted for each class in metric and/or imperial units as considered appropriate bearing in mind common UK usage.

The following abbreviations are used in class headings and also throughout this publication:

a.c.	alternating current.
BR	British Railways.
BSI	Bergische Stahl Industrie.
d.c.	direct current.
DEMU	Diesel-electric multiple unit.
Dia.	Diagram number.
DMU	Diesel multiple unit (general term).
EMU	Electric multiple unit.
hp	horsepower.
Hz	Hertz.
kN	kilonewtons.
km/h	kilometres per hour.
kW	kilowatts.
LT	London Transport.
LUL	London Underground Limited.
m.	metres.
mph	miles per hour.
RSL	Rolling Stock Library.
SR	Southern Railway.
t.	tonnes.
T	Toilets.

TD	Toilets suitable for disabled passengers.
V	volts.
W	Wheelchair spaces.

All dimensions and weights are quoted for vehicles in an 'as new' condition with all necessary supplies on board. Dimensions are quoted in the order Length – Width – Height. All lengths quoted are over buffers or couplings as appropriate. All width and height dimensions quoted are maxima. Height of vehicles equipped with pantographs is quoted with pantograph lowered.

Bogie Types are quoted in the format powered/non-powered (e.g BP20/BT13 denotes BP20 powered bogies and BT non-powered bogies).

DETAIL DIFFERENCES

Only detail differences which currently affect the areas and types of train which vehicles may work are shown. All other detail differences are specifically excluded. Where such differences occur within a class or part class, these are shown alongside the individual set or vehicle number. Meaning of abbreviations is detailed in individual class headings.

LIVERY CODES

Livery codes are used in this publication to denote the various liveries carried by vehicles. Readers should note it is impossible in a publication of this size to list every livery variation which currently exists. In particular items ignored for the purposes of this book include:

- Minor colour variations;
- Omission of logos;
- All numbering, lettering and branding.

The descriptions quoted are thus a general guide only and may be subject to slight variation between individual vehicles.

A complete list of livery codes used in this publication appears on page 94.

OWNER CODES

Owner codes are used in this publication to denote the owners of vehicles listed. Most vehicles are leased by the TOCs from specialist leasing companies.

A complete list of owner codes used in this publication appears on page 95.

OPERATION CODES

Operation codes are used in this publication to denote the normal usage of the vehicles listed – i.e. A guide to the services of which train operating company any vehicle will normally be used upon. Where vehicles are used for non revenue earning purposes, an indication to the normal type of usage is given in the class heading. Where no operation code is shown, vehicles are currently not in use.

A complete list of operation codes used in this publication appears on page 95.

DEPOT & LOCATION CODES

Depot codes are used to denote the normal maintenance base of each operational vehicle. However, maintenance may be carried out at other locations and may also be carried out by mobile maintenance teams.

Location codes are used to denote the current actual location of stored vehicles. A location code will be followed by (S) to denote stored.

A complate list of depot and location codes used in this publication appears on page 96.

SET FORMATIONS

Set formations shown are those normally maintained. Readers should note some set formations may be temporarily varied from time to time to suit maintenance and/or operational requirements. Vehicles shown as 'Spare' are not formed in any regular set formation.

NAMES

Only names carried with official sanction are listed in this publication. As far as possible names are shown in UPPER/lower case characters as actually shown on the name carried on the vehicle(s). Unless otherwise shown, officially complete units are regarded as named rather than just the individual car(s) which carry the name.

GENERAL INFORMATION

CLASSIFICATION AND NUMBERING

25 kV a.c. 50 Hz overhead and dual system EMUs are classified in the series 300–399.

750 V d.c. third rail EMUs are classified in the series 400–599.

Service units are classified in the series 900–949.

EMU individual cars are numbered in the series 61000–78999, except for vehicles used on the Isle of Wight – which are numbered in a separate series.

Prior to privatisation, Service Stock individual cars were numbered in the series 975000–975999 and 977000–977999, although this series was not used exclusively for EMU vehicles. Since privatisation, use of these series has been sporadic, vehicles often now retaining their former numbers.

DESIGN CONSIDERATIONS

Unless otherwise stated all vehicles listed have bar couplings at non-driving ends and tread brakes. In all types of vehicle except 'Express' stock, seating is 3 + 2 (i.e. three seats on one side of the gangway plus two on the other side) in standard class open vehicles, 2 + 2 in first class open vehicles, 8 per compartment in standard class and 6 per compartment in first class. In Express stock, seating is 2 + 2 in standard class open vehicles and 2 + 1 in first class open vehicles.

OPERATING CODES

These codes are used by TOC staff to describe the various different types of vehicles and normally appear on data panels on the non-driving ends of vehicles.

ATC	Auxiliary Equipment Trailer Composite
ATS	Auxiliary Equipment Trailer Standard
BDBS	Battery Driving Trailer Brake Standard
BDMS	Battery Driving Motor Standard
BDTC	Battery Driving Trailer Composite
BDTS	Battery Driving Trailer Standard
DM	Driving Motor
DMBS	Driving Motor Brake Standard
DMC	Driving Motor Composite
DMF	Driving Motor First
DMLF	Driving Motor Lounge First
DMLV	Driving Motor Luggage Van
DMS	Driving Motor Standard
DT	Driving Trailer
DTB	Driving Trailer Brake

DTBS	Driving Trailer Brake Standard
DTC	Driving Trailer Composite
DTF	Driving Trailer First
DTV	Driving Trailer Van
DTS	Driving Trailer Standard
M	Motor
MB	Motor Brake
MBLS	Motor Brake Restaurant Buffet Lounge Standard
MBS	Motor Brake Standard
MS	Motor Standard
PMB	Pantograph Motor Buffet Standard
PMS	Pantograph Motor Standard
PMV	Pantograph Motor Van
PTS	Pantograph Trailer Standard
TAV	Trailer Auxiliary Equipment Van
TBC	Trailer Brake Composite
TBF	Trailer Brake First
TBS	Trailer Brake Standard
TC	Trailer Composite
TF	Trailer First
TFH	Trailer First with Handbrake
TRBS	Trailer Restaurant Buffet Standard
TS	Trailer Standard
TSD	Trailer Standard with Disabled Persons' toilet.
TSH	Trailer Standard with Handbrake
TSW	Trailer Standard with Wheelchair Accommodation

The letters 'O' (denoting Open), 'K' (denoting Corridor) or 'L' (denoting Lavatory) may be added to the above codes on some vehicle data panels. Where two vehicles of the same type are formed within the same unit, the above codes may be suffixed by (A) and (B) to differentiate between the vehicles.

A composite is a vehicle containing both first and standard class accommodation, although first class accommodation on some EMU vehicles has now all been permanently declassified. A brake vehicle is a vehicle containing separate specific accommodation for the conductor (as opposed to the use of rear or intermediate cabs on some units).

Single motor coach 25 kV a.c. 50 Hz overhead EMUs (except Class 306) all have the pantograph mounted on the motor coach. Units with more than one motor coach have the pantograph mounted on a trailer car denoted as shown above.

DESIGN CODES AND DIAGRAM CODES

For each type of vehicle the Rolling Stock Library issues a seven character 'Design Code' consisting of two letters plus four numbers and a suffix letter. (e.g. EF2110A). The first five characters of the Design Code are known as the 'Diagram Code' and these are quoted in this publication in sub-headings. The meaning of the various characters of the Design Code is as follows:

First Character

E	Electric Multiple Unit
L	Eurostar Unit

Second Character (EMU vehicles)

A	Driving Motor
B	Driving Motor Brake
C	Non-driving Motor
D	Non-driving Motor Brake
E	Driving Trailer
F	Battery Driving Trailer
G	Driving Trailer Brake
H	Trailer
I	Battery Driving Motor
J	Trailer Brake
N	Trailer Buffet
O	Battery Driving Trailer Brake
P	Trailer with Handbrake
X	Driving Motor Van
Z	All types of service vehicle

Second Character (Eurostar vehicles)

A	Driving Motor
B	Non-driving Motor
C	Trailer (with train manager's compartment) – position 3
D	Trailer – position 4
E	Trailer (with public telephone) – position 5
F	Trailer – position 6
G	Kitchen/Bar
H	Trailer – position 8
J	Trailer (with public telephone) – position 9
K	Trailer (with staff compartment) – position 10

Third Character

1	First class accommodation
2	Standard class accommodation
3	Composite accommodation
4	Unclassified accommodation
5	No passenger accommodation

Fourth & Fifth Characters
These distinguish between different designs of vehicle, each design being allocated a unique two digit number.

Special Note
Where vehicles have been declassified, the correct design code for a declassified vehicle is quoted in this publication, even though this may be at variance with RSL records, which do not always show the reality of the current position.

ACCOMMODATION

The information given in class headings and sub-headings is in the form F/S nT (or TD) nW. For example 12/54 1T 1W denotes 12 first class and 54 standard class seats, 1 toilet and 1 wheelchair space.

BUILD DETAILS

Lot Numbers

Lot (batch) numbers allocated when ordered (where available) are quoted in class headings and sub-headings

Builders/Heavy Maintenance Providers

These are shown in class headings, the following abbreviations being used:

ABB Derby	ABB, Derby Carriage Works (later Adtranz Derby).
ABB York	ABB, York.
Adtranz Chart Leacon	DaimlerChrysler Rail Systems (UK), Ashford, Kent
Adtranz Derby	DaimlerChrysler Rail Systems (UK), Derby.
Adtranz Ilford	DaimlerChrysler Rail Systems (UK), Ilford.
Alstom Birmingham	Alstom, Saltley, Birmingham.
Alstom Eastleigh	Alstom, Eastleigh.
Ashford	BR, Ashford Works.
BRCW	Birmingham Railway Carriage & Wagon, Smethwick.
BREL Derby	BREL, Derby Carriage Works (later ABB Derby).
BREL York	BREL, York (later ABB York).
BREL Eastleigh	BREL, Eastleigh(later Wessex Traincare).
BREL Swindon	BREL, Swindon.
BREL Wolverton	BREL, Wolverton (later Railcare, Wolverton).
CAF	Construcciones y Auxiliar de Ferrocarriles, Zaragosa, Spain.
Cravens	Cravens, Sheffield.
Derby	BR, Derby Carriage Works (later BREL Derby).
Doncaster	BR, Doncaster Works (later BREL Doncaster)
Eastleigh	BR, Eastleigh Works (later BREL Eastleigh)
GEC-Alsthom B'ham	GEC-Alsthom, Saltley, Birmingham (later Alstom Birmingham)
Hunslet TPL	Hunslet Transportation Projects, Leeds.
Met-Camm.	Metropolitan-Cammell, Saltley, Birmingham (later GEC-A Birmingham).
Pressed Steel	Pressed Steel, Linwood.
Railcare Glasgow	Railcare, Springburn, Glasgow.
Railcare Wolverton	Railcare, Wolverton.
Wessex Traincare	Wessex Traincare, Eastleigh (later Alstom Eastleigh).
York	BR, York Carriage Works (later BREL York)

The previous practice of showing dual builder details (e.g. Ashford/Eastleigh) where vehicle underframe and body were built at two separate locations has been discontinued as this is now virtually the industry norm rather than an exceptional circumstance.

ELECTRIC MULTIPLE UNITS

1. 25 kV a.c. 50 Hz OVERHEAD UNITS

Supply System: Except where otherwise stated, all units in this section operate on 25 kV a.c. 50 Hz overhead only.

CLASS 303 3-Car Unit

DTS–MBS–BDTS. Gangwayed within unit. 2 + 2 (* 2 + 3) seating.
Traction Motors: Four Metropolitan Vickers MV155 of 155 kW each.
Dimensions: 20.18 x 2.82 x 3.86 m.
Maximum Speed: 75 mph. **Doors**: Power operated sliding.
Couplings: Buckeye. **Bogies**: Gresley ED3/ET3.
Multiple Working: Classes 303–312 only.

61481–61515. MBS. Dia. ED220. Lot No. 30580 Pressed Steel 1959–60. –/48. 56.4 t.
61812–61867. MBS. Dia. ED220. Lot No. 30630 Pressed Steel 1960–61. –/48. 56.4 t.
75566–75600. DTS. Dia. EE241. Lot No. 30579 Pressed Steel 1959–60. –/56. 34.4 t.
75601–75635. BDTS. Dia. EF217. Lot No. 30581 Pressed Steel 1959–60. –/56. 38.4 t.
75746–75801. DTS. Dia. EE241 (* EE206). Lot No. 30629 Pressed Steel 1960–61. –/56 (* –/83). 34.4 (* 34.5) t.
75802–75857. BDTS. Dia. EF217. Lot No. 30631 Pressed Steel 1960–61. –/56. 38.4 t.

303 001	**S**	A	*SR*	GW	75566	61481	75601
303 003	**S**	A	*SR*	GW	75568	61483	75603
303 004	**S**	A	*SR*	GW	75569	61484	75604
303 006	**S**	A	*SR*	GW	75571	61486	75606
303 008	**S**	A	*SR*	GW	75573	61488	75608
303 009	**S**	A	*SR*	GW	75574	61489	75609
303 010	**S**	A	*SR*	GW	75575	61490	75610
303 011	**S**	A	*SR*	GW	75576	61491	75611
303 012	**S**	A	*SR*	GW	75577	61492	75612
303 013	**S**	A	*SR*	GW	75578	61493	75613
303 014	**S**	A	*SR*	GW	75579	61494	75614
303 016	**S**	A	*SR*	GW	75750	61496	75616
303 019	**CC**	A	*SR*	GW	75584	61499	75619
303 020	**S**	A	*SR*	GW	75585	61500	75620
303 021	**CC**	A	*SR*	GW	75586	61501	75621
303 023	**CC**	A	*SR*	GW	75588	61503	75623
303 024	**S**	A		GW(S)	75589	61504	75624
303 025	**S**	A	*SR*	GW	75590	61505	75625

303 027	**S**	A	*SR*	GW	75592	61507	75627
303 028	**S**	A		GW(S)	75600	61813	75845
303 032	**S**	A	*SR*	GW	75597	61512	75632
303 033	**S**	A	*SR*	GW	75595	61860	75817
303 034	**S**	A	*SR*	GW	75599	61514	75634
303 037	**S**	A	*SR*	GW	75781	61508	75803
303 040	**S**	A	*SR*	GW	75581	61816	75806
303 043	**S**	A	*SR*	GW	75572	61819	75809
303 045	**S**	A	*SR*	GW	75755	61821	75811
303 047	**S**	A	*SR*	GW	75757	61823	75813
303 054	**S**	A	*SR*	GW	75764	61830	75820
303 055	**S**	A	*SR*	GW	75765	61831	75821
303 056	**S**	A	*SR*	GW	75766	61832	75822
303 058	**S**	A	*SR*	GW	75768	61834	75824
303 061	**S**	A	*SR*	GW	75771	61837	75827
303 065	**S**	A	*SR*	GW	75775	61841	75831
303 070	**S**	A	*SR*	GW	75780	61846	75836
303 077	**S**	A	*SR*	GW	75787	61853	75843
303 079	**S**	A	*SR*	GW	75789	61855	75635
303 080	**S**	A	*SR*	GW	75790	61856	75846
303 083	**S**	A	*SR*	GW	75793	61859	75849
303 085	**S**	A	*SR*	GW	75795	61861	75851
303 087	**CC**	A	*SR*	GW	75797	61863	75853
303 088	**S**	A	*SR*	GW	75798	61864	75854
303 089	**S**	A	*SR*	GW	75799	61865	75855
303 090	**S**	A	*SR*	GW	75800	61866	75856
303 091	**S**	A	*SR*	GW	75801	61867	75857
Spare *	**BG**	A		LT(S)	75773		

Name (carried on MBS):

303 089 COWAL HIGHLAND GATHERING 1894–1994

CLASS 305 — 3- or 4-Car Unit

BDTC (declassified)–MBS–DTS or BDTS–MBS–TS–DTS. Gangwayed within unit.
Traction Motors: Four GEC WT380 of 153 kW each.
Dimensions: 20.35 (BDTC, BDTS & DTS) or 20.29 (MBS & TS) x 2.82 x 3.84 m.
Maximum Speed: 75 mph. **Doors**: Manually operated slam.
Couplings: Buckeye. **Bogies**: Gresley ED5/ET5.
Multiple Working: Classes 303–312 only.

61410–61428. MBS. Dia. ED216. Lot No. 30567 Doncaster 1960. –/76 (*–/58; †–/72). 56.5 t.
70356–70374. TS. Dia. EH223. Lot No. 30568 Doncaster 1960. –/86 1T. 31.5 t.
75424–75442. BDTC († BDTS). Dia. EF304 († EF2??). Lot No. 30566 Doncaster 1960. 24/52 (*20/40; †–/80) 1T. 36.5 t.
75443–75461. DTS. Dia. EE220. Lot No. 30569 Doncaster 1960. –/88 (* –/70). 32.7 t.

305 501	†	**RR**	A	*SR*	GW	75424	61410	70356	75443
305 502	†	**RR**	A	*SR*	GW	75425	61421	70357	75444
305 506		**GM**	A	*NW*	LG	75429	61415		75448
305 508	†	**RR**	A	*SR*	GW	75431	61417	70363	75450
305 511	*	**GM**	A	*NW*	LG	75434	61420		75453
305 516		**GM**	A		LG(S)	75439	61425		75458
305 517	†	**RR**	A	*SR*	GW	75440	61426	70372	75459
305 519	†	**RR**	A	*SR*	GW	75442	61428	70374	75461

CLASS 306 — 3-Car Unit

DMS–TBS–DTS. Non gangwayed. 2 + 2 seating. Original supply system 1500 V d.c. overhead, converted 1960–61. Retained for special workings only, not used in normal service.
Traction Motors: Four Crompton Parkinson of 155 kW each.
Dimensions: 19.24 (DMS & DTS) or 17.40 (TBS) x 2.89 x 3.84 m.
Maximum Speed: 70 mph. **Doors**: Power operated sliding.
Couplings: Screw. **Bogies**: LNER ED6/ET6.
Multiple Working: Classes 303–312 only.

65201–65292. DMS. Dia. EA203. Lot No. 363 Met-Camm. 1949. –/62. 51.7 t.
65401–65492. TBS. Dia. EJ201. Lot No. 365 BRCW 1949. –/46. 26.4 t.
65601–65692. DTS. Dia. EE211. Lot No. 364 Met-Camm. 1949. –/60. 27.9 t.

306 017	**G**	H		IL(S)	65217	65417	65617

CLASS 308 — 3-Car Unit

BDTC (declassified)–MBS–DTS. Gangwayed within unit. Originally 4-car units, but surviving TS cars are all now spare.
Traction Motors: Four English Electric 536A of 143.5 kW each.
Dimensions: 19.88 (BDTC & DTS) or 19.35 (MBS & TS) x 2.82 x 3.86 m.
Maximum Speed: 75 mph. **Doors**: Manually operated slam.
Couplings: Buckeye. **Bogies**: Gresley ED5/ET5.
Multiple Working: Classes 303–312 only.

61883–61891. MBS. Dia. ED216. Lot No. 30653 York 1961–62. –/76. 55.0 t.
61893–61915. MBS. Dia. ED216. Lot No. 30657 York 1961–62. –/76. 55.0 t.
70611–70619. TS. Dia. EH223. Lot No. 30654 York 1961. Converted from TC. –/86 1T. 30.0 t.
70620–70643. TS. Dia. EH223. Lot No. 30658 York 1961–62. Converted from TC. –/86 1T. 30.0 t.
75878–75886. BDTC. Dia. EF304. Lot No. 30652 York 1961–62. 24/50 1T. 36.3 t.
75887–75895. DTS. Dia. EE220. Lot No. 30655 York 1961–62. –/88. 33.0 t.
75896–75928. BDTC. Dia. EF304. Lot No. 30656 York 1961–62. 24/52 1T. 36.3 t.
75929–75961. DTS. Dia. EE220. Lot No. 30659 York 1961–62. –/88. 33.0 t.

308 134	**WY**	A		ZH(S)	75879	61884	75888
308 136	**WY**	A	*NS*	NL	75881	61886	75890
308 137	**WY**	A	*NS*	NL	75882	61887	75891
308 138	**WY**	A	*NS*	NL	75883	61888	75892
308 141	**WY**	A	*NS*	NL	75886	61891	75895

308 143	**WY**	A	*NS*	NL	75897	61893	75930	
308 144	**WY**	A	*NS*	NL	75880	61894	75931	
308 145	**WY**	A	*NS*	NL	75899	61895	75932	
308 147	**WY**	A	*NS*	NL	75901	61897	75934	
308 152	**WY**	A	*NS*	NL	75913	61902	75939	
308 153	**WY**	A	*NS*	NL	75907	61903	75940	
308 154	**WY**	A	*NS*	NL	75908	61904	75941	
308 155	**WY**	A	*NS*	NL	75909	61905	75942	
308 157	**WY**	A	*NS*	NL	75915	61907	75944	
308 158	**WY**	A	*NS*	NL	75912	61908	75945	
308 159	**WY**	A	*NS*	NL	75906	61909	75946	
308 161	**WY**	A	*NS*	NL	75911	61911	75948	
308 162	**WY**	A	*NS*	NL	75916	61912	75949	
308 163	**WY**	A	*NS*	NL	75917	61913	75950	
308 164	**WY**	A	*NS*	NL	75918	61914	75951	
308 165	**WY**	A	*NS*	NL	75919	61915	75952	

Spare TS.

Spare	**N**	A		CB(S)	70612	70621	70622	70631
Spare	**N**	A		CB(S)	70640			

CLASS 309 'CLACTON' 4-Car Express Unit

BDTC–MBS–TS–DTS. Gangwayed throughout.
Traction Motors: Four GEC WT401 of 210 kW each.
Dimensions: 20.18 x 2.82 x 3.90 m.
Maximum Speed: 100 mph.
Doors: Manually operated slam.
Couplings: Buckeye.
Bogies: Commonwealth.
Multiple Working: Classes 303–312 only.
Advertising Livery:

- 309 624 'Manchester Airport Air Express'.

61925–61931. MBS. Dia. ED218. Lot No. 30676 York 1962. –/52. 57.7 t.
61932–61939. MBS. Dia. ED218. Lot No. 30680 York 1962. –/52. 57.7 t.
70253–70259. TS. Dia. EH229. Lot No. 30677 York 1962. –/68. 34.8 t.
71754–71761. TS. Dia. EH228. Lot No. 31001 BREL Wolverton 1984–87. Converted from loco-hauled vehicles built to Lot No. 30724 York 1962–63. –/68. 34.8 t.
75637–75644. BDTC. Dia. EF305. Lot No. 30679 York 1962. 18/32 2T. 40.0 t.
75962–75968. BDTC. Dia. EF305. Lot No. 30675 York 1962. 18/32 2T. 40.0 t.
75969–75975. DTS. Dia. EE229. Lot No. 30678 York 1962. –/56 2T. 36.6 t.
75976–75983. DTS. Dia. EE229. Lot No. 30682 York 1962–63. –/56 2T. 36.6 t.

309 613	**RN**	A	*NW*	LG	75639	61934	71756	75978
309 616	**RN**	A	*NW*	LG	75642	61937	71759	75981
309 617	**RN**	A	*NW*	LG	75643	61938	71760	75982
309 623	**RN**	A	*NW*	LG	75641	61927	71758	75980
309 624	**AL**	A	*NW*	LG	75965	61928	70256	75972
309 627	**RN**	A	*NW*	LG	75644	61931	70259	75975

▲ 303 045 arrives at Glasgow Central on 12th October 1999 with a service from Gourock. This class is scheduled for replacement by Class 334 early in 2000.
Bob Sweet

▼ 310 089 passes Shadwell station of the Docklands Light Railway with the 10.10 London Fenchurch Street–Shoeburyness on 30th May 1998. **Kevin Conkey**

West Anglia Great Northern operated 313 060 wears an attractive advertising livery for the public transport initiative 'Intalink' as an alternative to the white undercoat livery worn by most other refurbished members of this class pending adoption of a new livery.
Brian Morrision

▲ 314 213 departs from Glasgow Central with a Cathcart Circle service on 12th October 1999. **Bob Sweet**

▼ 315 816 nears Bow Junction, Stratford, with a service for London Liverpool Street on 9th September 1999. **Hugh Ballantyne**

▲ Conversion of Class 317/2 to Class 317/6 was completed during 1999. 317 670 was photographed at Hackney Downs with the 13.42 Bishop's Stortford–London Liverpool Street on 26th May 1999. **Kevin Conkey**

▼ 318 253 passes Troon Golf Course, Ayrshire, with an Ayr–Glasgow Central working on 28th April 1999. **Paul Senior**

▲ The 10.08 London Victoria–Brighton service departs from Clapham Junction on 22nd March 1997 formed of 319 220 and 319 218. **Kevin Conkey**

▼ 320 319 arrives at Springburn with an empty working from Yoker SD on 13th August 1999. **Hugh Ballantyne**

321 436 passes Headstone Lane whilst working the 15.34 London Euston–Milton Keynes Central on 1st April 1999.
David Brown

▲ 322 483 passes through Bethnal Green with the 10.30 London Liverpool Street–Stansted Airport on 16th May 1999. These units are to be replaced on this service by Class 317/7 early in 2000. **Kevin Conkey**

▼ 323 223, one of three units with revised seating layout for use on services to/from Manchester Airport, leaves Stoke on Trent with the 09.08 to Manchester Piccadilly on 6th September 1999. **Colin J. Marsden**

▲ 325 005 passing Wandsworth Road with 1M44, 13.47 Tonbridge Royal Mail Terminal–Willesden Royal Mail Terminal on 6th August 1998. **Hugh Ballantyne**

▼ Heathrow Express unit 332 007 approaches London Paddington on 16th June 1998. **John G. Teasdale**

CLASS 310 3- or 4-Car Unit

Various formations, see below. Gangwayed within unit. Disc brakes.
Traction Motors: Four English Electric 546 of 201.5 kW each.
Dimensions: 20.18 x 2.82 x 3.86 m.
Maximum Speed: 75 mph. **Doors:** Manually operated slam.
Couplings: Buckeye. **Bogies:** B4.
Multiple Working: Classes 303–312 only.
Note: 310 081 carries car numbers 76991, 62526, 71210, 78042 in error on one side only.

62071–62120. MBS. Dia. ED219. Lot No. 30746 Derby 1965–67. –/68. 57.2 t.
70731–70780. TS. Dia. EH232. Lot No. 30747 Derby 1965–67. –/98. 31.7 t.
76130–76179. BDTS. Dia. EF211. Lot No. 30745 Derby 1965–67. –/80 2T. 37.3 t.
76180/181/183–186/190–195/198–205/208/209/211/213–223/225/76227/229. DTC. Dia. EE306. Lot No. 30748 Derby 1965–67. 25/43 2T. 34.4 t.
76182/187–189/196/197/206/207/210/212/224. DTS. Dia. EE237. Lot No. 30748 Derby 1965–67. Converted from DTC. –/75 2T. 34.4 t.
76228. BDTS. Dia. EF210. Lot No. 30748 Derby 1967. Converted from DTC. –/68 2T. 34.5 t.
76998. BDTS. Dia. EF214. Lot No. 30747 Derby 1965–67. Converted from TS. –/75 2T. 35.0 t.

Class 310/0. 4-car units. BDTS–MBS–TS–DTC (declassified).

310 046	**N**	H	*LS*	EM	76130	62071	70731	76180
310 047	**N**	H	*LS*	EM	76131	62072	70732	76181
310 049	**N**	H	*LS*	EM	76133	62074	70734	76183
310 050	**N**	H	*LS*	EM	76134	62075	70735	76184
310 051	**N**	H	*LS*	EM	76135	62076	70736	76185
310 052	**N**	H	*LS*	EM	76136	62077	70737	76186
310 057	**N**	H	*LS*	EM	76141	62082	70742	76191
310 058	**N**	H	*LS*	EM	76142	62083	70743	76192
310 059	**N**	H	*LS*	EM	76143	62084	70744	76205
310 060	**N**	H	*LS*	EM	76144	62085	70745	76194
310 064	**N**	H	*LS*	EM	76148	62089	70749	76198
310 066	**N**	H	*LS*	EM	76228	62091	70751	76200
310 067	**N**	H	*LS*	EM	76151	62092	70752	76201
310 068	**N**	H	*LS*	EM	76152	62093	70753	76202
310 069	**N**	H	*LS*	EM	76153	62094	70754	76203
310 070	**N**	H	*LS*	EM	76154	62095	70755	76204
310 074	**N**	H	*LS*	EM	76145	62099	70759	76208
310 075	**N**	H	*LS*	EM	76159	62100	70760	76209
310 077	**N**	H	*LS*	EM	76161	62102	70762	76211
310 079	**N**	H	*LS*	EM	76163	62104	70764	76222
310 080	**N**	H	*LS*	EM	76164	62105	70765	76214
310 081	**N**	H	*LS*	EM	76165	62106	70766	76215
310 082	**N**	H	*LS*	EM	76166	62107	70767	76216
310 083	**N**	H	*LS*	EM	76167	62108	70768	76217
310 084	**N**	H	*LS*	EM	76168	62109	70769	76218
310 085	**N**	H	*LS*	EM	76169	62110	70770	76219

310 086	**N**	H	*LS*	EM	76170	62111	70771	76220
310 087	**N**	H	*LS*	EM	76171	62112	70772	76221
310 088	**N**	H	*LS*	EM	76172	62113	70773	76213
310 089	**N**	H	*LS*	EM	76173	62114	70774	76223
310 091	**N**	H	*LS*	EM	76175	62116	70776	76225
310 092	**N**	H	*LS*	EM	76176	62117	70777	76226
310 093	**N**	H	*LS*	EM	76177	62118	70778	76190
310 094	**N**	H	*LS*	EM	76998	62119	70780	76193
310 095	**N**	H	*LS*	EM	76179	62120	70779	76229

Name (carried on MBS):

310 058 Chafford Hundred.

Class 310/1. 3-car units. BDTS–MBS–DTS (*DTC (declassified)).

310 101		**RR**	H		EM(S)	76157	62098	76207
310 102		**RR**	H	*LS*	EM	76139	62080	76189
310 103		**RR**	H		EM(S)	76160	62101	76210
310 104		**RR**	H	*LS*	EM	76162	62103	76212
310 105		**RR**	H	*CT*	SI	76174	62115	76224
310 106		**RR**	H	*LS*	EM	76156	62097	76206
310 107		**RR**	H	*CT*	SI	76146	62087	76196
310 108		**RR**	H	*CT*	SI	76132	62073	76182
310 109		**RR**	H		EM(S)	76137	62078	76187
310 110		**RR**	H	*LS*	EM	76138	62079	76188
310 111		**RR**	H	*CT*	SI	76147	62088	76197
310 112	*	**RR**	H	*CT*	SI	76140	62086	76227
310 113	*	**RR**	H	*CT*	SI	76158	62090	76195

Spare TS.

Spare	**PM**	H	KN(S)	70733	70747	70748	70757
Spare	**PM**	H	KN(S)	70763			

CLASS 312 4-Car Unit

BDTS–MBS–TS–DTC. Gangwayed within unit. Disc brakes. All DTC vehicles operated by LTS Rail are declassified.
Traction Motors: Four English Electric 546 of 201.5 kW each.
Dimensions: 20.18 x 2.82 x 3.86 m.
Maximum Speed: 90 mph. **Doors:** Manually operated slam.
Couplings: Buckeye. **Bogies:** B4.
Multiple Working: Classes 303–312 only.
Note: 76949 carries number 76946 in error on one side only.

Class 312/0. Built to operate on 25 kV 50 Hz a.c. overhead only.

62484–62509. MBS. Dia. ED212. Lot No. 30864 BREL York 1977–78. –/68. 56.0 t.
62657–62560. MBS. Dia. ED214. Lot No. 30892 BREL York 1976. –/68. 56.0 t.
71168–71193. TS. Dia. EH209. Lot No. 30865 BREL York 1977–78. –/98. 30.5 t.
71277–71280. TS. Dia. EH209. Lot No. 30893 BREL York 1976. –/98. 30.5 t.
76949–76974. BDTS. Dia. EF213. Lot No. 30863 BREL York 1977–78. –/84 1T. 34.9 t.
76994–76997. BDTS. Dia. EF213. Lot No. 30891 BREL York 1976. –/84 1T. 34.9 t.

78000–78025. DTC. Dia. EE305. Lot No. 30866 BREL York 1977–78. 25/47 2T. 33.0 t.
78045–78048. DTC. Dia. EE305. Lot No. 30894 BREL York 1976. 25/47 2T. 33.0 t.

312 701	**GE**	A	*GE*	IL	76949	62484	71168	78000
312 702	**GE**	A	*GE*	IL	76950	62485	71169	78001
312 703	**GE**	A	*GE*	IL	76951	62486	71170	78002
312 704	**GE**	A	*LS*	EM	76952	62487	71171	78003
312 705	**GE**	A	*GE*	IL	76953	62488	71172	78004
312 706	**GE**	A	*GE*	IL	76954	62489	71173	78005
312 707	**GE**	A	*GE*	IL	76955	62490	71174	78006
312 708	**GE**	A	*LS*	EM	76956	62491	71175	78007
312 709	**GE**	A	*GE*	IL	76957	62492	71176	78008
312 710	**GE**	A	*GE*	IL	76958	62493	71177	78009
312 711	**GE**	A	*GE*	IL	76959	62494	71178	78010
312 712	**GE**	A	*GE*	IL	76960	62495	71179	78011
312 713	**GE**	A	*GE*	IL	76961	62496	71180	78012
312 714	**GE**	A	*GE*	IL	76962	62497	71181	78013
312 715	**GE**	A	*GE*	IL	76963	62498	71182	78014
312 716	**GE**	A	*GE*	IL	76964	62499	71183	78015
312 717	**GE**	A	*GE*	IL	76965	62500	71184	78016
312 718	**GE**	A	*GE*	IL	76966	62501	71185	78017
312 719	**GE**	A	*GE*	IL	76967	62502	71186	78018
312 720	**GE**	A	*GE*	IL	76968	62503	71187	78019
312 721	**GE**	A	*GE*	IL	76969	62504	71188	78020
312 722	**GE**	A	*GE*	IL	76970	62505	71189	78021
312 723	**GE**	A	*GE*	IL	76971	62506	71190	78022
312 724	**GE**	A	*GE*	IL	76972	62507	71191	78023
312 725	**N**	A	*LS*	EM	76973	62509	71193	78025
312 726	**N**	A	*LS*	EM	76974	62508	71192	78024
312 727	**N**	A	*LS*	EM	76994	62657	71277	78045
312 728	**N**	A	*LS*	EM	76995	62658	71278	78046
312 729	**N**	A	*LS*	EM	76996	62659	71279	78047
312 730	**N**	A	*LS*	EM	76997	62660	71280	78048

Class 312/1. Built to operate on 25 kV 50 Hz a.c. or 6.25 kV 50 Hz a.c. overhead.

62510–62528. MBS. Dia. ED213. Lot No. 30868 BREL York 1975–76. –/68. 56.0 t.
71194–71212. TS. Dia. EH209. Lot No. 30869 BREL York 1975–76. –/98. 30.5 t.
76975–76993. BDTS. Dia. EF213. Lot No. 30867 BREL York 1975–76. –/84 2T. 34.9 t.
78026–78044. DTC. Dia. EE305. Lot No. 30870 BREL York 1975–76. 25/47 2T. 33.0 t.

312 781	**N**	A	*LS*	EM	76975	62510	71194	78026
312 782	**N**	A	*LS*	EM	76976	62511	71195	78027
312 783	**N**	A	*LS*	EM	76977	62512	71196	78028
312 784	**N**	A	*LS*	EM	76978	62513	71197	78029
312 785	**N**	A	*LS*	EM	76979	62514	71198	78030
312 786	**N**	A	*LS*	EM	76980	62515	71199	78031
312 787	**N**	A	*LS*	EM	76981	62516	71200	78032
312 788	**N**	A	*LS*	EM	76982	62517	71201	78033
312 789	**N**	A	*LS*	EM	76983	62518	71202	78034

312 790	**N**	A	*LS*	EM	76984	62519	71203	78035
312 791	**N**	A	*LS*	EM	76985	62520	71204	78036
312 792	**N**	A	*LS*	EM	76986	62521	71205	78037
312 793	**N**	A	*LS*	EM	76987	62522	71206	78038
312 794	**N**	A	*LS*	EM	76988	62523	71207	78039
312 795	**N**	A	*LS*	EM	76989	62524	71208	78040
312 796	**N**	A	*LS*	EM	76990	62525	71209	78041
312 797	**N**	A	*LS*	EM	76991	62526	71210	78042
312 798	**N**	A	*LS*	EM	76992	62527	71211	78043
312 799	**N**	A	*LS*	EM	76993	62528	71212	78044

CLASS 313 — 3-Car Unit

DMS–PTS–BDMS. Gangwayed within unit. End doors. Disc and rheostatic brakes.
Supply System: 25 kV 50 Hz a.c. overhead or 750 V d.c. third rail.
Traction Motors: Four GEC G310AZ of 82.125 kW each.
Dimensions: 19.80 (DMS & BDMS) or 19.92 (PTS) x 2.82 x 3.58 m.
Maximum Speed: 75 mph. **Doors:** Power operated sliding.
Couplers: Tightlock. **Bogies:** BREL BX1.
Multiple Working: Classes 313–323.
Notes: 71217 carries number 71277 in error on one side only. 71733[II] is actually 71233.

62529–62592. DMS. Dia. EA204. Lot No. 30879 BREL York 1976–77. –/74. 37.0 t.
62593–62656. BDMS. Dia. EI201. Lot No. 30885 BREL York 1976–77. –/74. 37.6 t.
71213–71276. PTS. Dia. EH210. Lot No. 30880 BREL York 1976–77. –/84 (*–/80). 31.0 t.

Class 313/0. Operated by West Anglia Great Northern Railway.

Note: † Refurbished 1998 onwards by Adtranz Ilford with high back seats.
Advertising Liveries:
- 313 043/057 'WAGN Family Travelcard'
- 313 060 'Intalink'

313 018		**N**	H	*WN*	HE	62546	71230	62610
313 024	†	**U**	H	*WN*	HE	62552	71236	62616
313 025		**N**	H	*WN*	HE	62553	71237	62617
313 026		**N**	H	*WN*	HE	62554	71238	62618
313 027	†	**U**	H	*WN*	HE	62555	71239	62619
313 028	†	**U**	H	*WN*	HE	62556	71240	62620
313 029		**N**	H	*WN*	HE	62557	71241	62621
313 030		**N**	H	*WN*	HE	62558	71242	62622
313 031		**N**	H	*WN*	HE	62559	71243	62623
313 032		**N**	H	*WN*	HE	62560	71244	62643
313 033		**N**	H	*WN*	HE	62561	71245	62625
313 035		**N**	H	*WN*	HE	62563	71247	62627
313 036		**N**	H	*WN*	HE	62564	71248	62628
313 037		**N**	H	*WN*	HE	62565	71249	62629
313 038		**N**	H	*WN*	HE	62566	71250	62630
313 039		**N**	H	*WN*	HE	62567	71251	62631

313 040		**N**	H	*WN*	HE	62568	71252	62632
313 041		**N**	H	*WN*	HE	62569	71253	62633
313 042		**N**	H	*WN*	HE	62570	71254	62634
313 043	†	**AL**	H	*WN*	HE	62571	71255	62635
313 044		**N**	H	*WN*	HE	62572	71256	62636
313 045		**N**	H	*WN*	HE	62573	71257	62637
313 046		**N**	H	*WN*	HE	62574	71258	62638
313 047		**N**	H	*WN*	HE	62575	71259	62639
313 048		**N**	H	*WN*	HE	62576	71260	62640
313 049		**N**	H	*WN*	HE	62577	71261	62641
313 050	†	**U**	H	*WN*	HE	62578	71262	62649
313 051		**N**	H	*WN*	HE	62579	71263	62624
313 052		**N**	H	*WN*	HE	62580	71264	62644
313 053		**N**	H	*WN*	HE	62581	71265	62645
313 054		**N**	H	*WN*	HE	62582	71266	62646
313 055		**N**	H	*WN*	HE	62583	71267	62647
313 056		**N**	H	*WN*	HE	62584	71268	62648
313 057	†	**AL**	H	*WN*	HE	62585	71269	62642
313 058		**N**	H	*WN*	HE	62586	71270	62650
313 059		**N**	H	*WN*	HE	62587	71271	62651
313 060	†	**AL**	H	*WN*	HE	62588	71272	62652
313 061		**N**	H	*WN*	HE	62589	71273	62653
313 062		**N**	H	*WN*	HE	62590	71274	62654
313 063		**N**	H	*WN*	HE	62591	71275	62655
313 064	†	**U**	H	*WN*	HE	62592	71276	62656

Class 313/1. Operated by Silverlink Train Services. Equipped with additional shoegear for working on LUL 750 V d.c. 4-rail system. Units are renumbered and reclassified from Class 313/0 upon completion of facelift by Railcare Wolverton, retaining last two digits of previous number.

Number	*Former No.*								
313 101	313 001	*	**SL**	H	SL	BY	62529	71213	62593
313 102	313 002	*	**SL**	H	SL	BY	62530	71214	62594
313 103	313 003	*	**SL**	H	SL	BY	62531	71215	62595
	313 004		**N**	H	SL	BY	62532	71216	62596
313 105	313 005	*	**SL**	H	SL	BY	62533	71217	62597
	313 006		**N**	H	SL	BY	62534	71218	62598
313 107	313 007	*	**SL**	H	SL	BY	62535	71219	62599
	313 008		**N**	H	SL	BY	62536	71220	62600
	313 009		**N**	H	SL	BY	62537	71221	62601
313 110	313 010	*	**SL**	H	SL	BY	62538	71222	62602
	313 011		**N**	H	SL	BY	62539	71223	62603
313 112	313 012	*	**SL**	H	SL	BY	62540	71224	62604
	313 013		**N**	H	SL	BY	62541	71225	62605
313 114	313 014	*	**SL**	H	SL	BY	62542	71226	62606
313 115	313 015	*	**SL**	H	SL	BY	62543	71227	62607
	313 016		**N**	H	SL	BY	62544	71228	62608
313 117	313 017	*	**SL**	H	SL	BY	62545	71229	62609
	313 019		**N**	H	SL	BY	62547	71231	62611
313 120	313 020	*	**SL**	H	SL	BY	62548	71232	62612
313 121	313 021	*	**SL**	H	SL	BY	62549	71733‖	62613

	313 022	**SL**	H	SL	BY	62550	71234	62614
	313 023	**N**	H	SL	BY	62551	71235	62615
313 134	313 034 *	**SL**	H	SL	BY	62562	71246	62626

Name (carried on PTS):

313 120 PARLIAMENT HILL

CLASS 314 3-Car Unit

DMS–PTS–DMS. Gangwayed within unit. End doors. Disc and rheostatic brakes.
Traction Motors: Four Brush TM61-53 or GEC G310AZ of 82.125 kW each.
Dimensions: 19.80 (DMS or 19.92 (PTS) x 2.82 x 3.58 m.
Maximum Speed: 75 mph. **Doors:** Power operated sliding.
Couplers: Tightlock. **Bogies:** BREL BX1.
Multiple Working: Classes 313–323.

64583–64613 (Odd numbers). DMS(A). Dia. EA206. Lot No. 30912 BREL York 1979. –/68. 34.5 t.
64584–64614 (Even numbers). DMS(B). Dia. EA206. Lot No. 30912 BREL York 1979. –/68. 34.5 t.
64588[II]. DMS(B). Dia. EA206. Lot No. 30908 BREL York 1978–80. Rebuilt Railcare Glasgow 1996 from Class 507 DMS. –/74. 35.6 t.
71450–71465. PTS. Dia. EH211. Lot No. 30913 BREL York 1979. –/76. 33.0 t.

Units 314 201–314 206. Brush traction motors.

314 201	**S**	A	*SR*	GW	64583	71450	64584
314 202	**S**	A	*SR*	GW	64585	71451	64586
314 203	**S**	A	*SR*	GW	64587	71452	64588[II]
314 204	**CC**	A	*SR*	GW	64589	71453	64590
314 205	**CC**	A	*SR*	GW	64591	71454	64592
314 206	**CC**	A	*SR*	GW	64593	71455	64594

Units 314 207–314 216. GEC traction motors.

314 207	**S**	A	*SR*	GW	64595	71456	64596
314 208	**S**	A	*SR*	GW	64597	71457	64598
314 209	**S**	A	*SR*	GW	64599	71458	64600
314 210	**CC**	A	*SR*	GW	64601	71459	64602
314 211	**CC**	A	*SR*	GW	64603	71460	64604
314 212	**S**	A	*SR*	GW	64605	71461	64606
314 213	**S**	A	*SR*	GW	64607	71462	64608
314 214	**CC**	A	*SR*	GW	64609	71463	64610
314 215	**CC**	A	*SR*	GW	64611	71464	64612
314 216	**CC**	A	*SR*	GW	64613	71465	64614

Name (carried on PTS):

314 203 European Union

CLASS 315 4-Car Unit

DMS–TS–PTS–DMS. Gangwayed within unit. End doors. Disc and rheostatic brakes.
Traction Motors: Four GEC G310AZ or Brush TM61-53) of 82.125 kW each.
Dimensions: 19.80 (DMS) or 19.92 (TS & PTS) x 2.82 x 3.58 m.
Maximum Speed: 75 mph. **Doors**: Power operated sliding.
Couplers: Tightlock. **Bogies**: BREL BX1.
Multiple Working: Classes 313–323.
Advertising Livery:
- 315 844 'WAGN Family Travelcard'.

64461–64581 (Odd numbers). DMS(A). Dia. EA207. Lot No. 30902 BREL York 1980–81. –/74. 35.0 t.
64462–64582 (Even numbers). DMS(B). Dia. EA207. Lot No. 30902 BREL York 1980–81. –/74. 35.0 t.
71281–71341. TS. Dia. EH216. Lot No. 30904 BREL York 1980–81. –/86. 25.5 t.
71389–71449. PTS. Dia. EH217. Lot No. 30903 BREL York 1980–81. –/84. 32.0 t.

Units 315 801–315 841. GEC traction motors.

315 801	**GE**	H	*GE*	IL	64461	71281	71389	64462
315 802	**GE**	H	*GE*	IL	64463	71282	71390	64464
315 803	**GE**	H	*GE*	IL	64465	71283	71391	64466
315 804	**GE**	H	*GE*	IL	64467	71284	71392	64468
315 805	**GE**	H	*GE*	IL	64469	71285	71393	64470
315 806	**GE**	H	*GE*	IL	64471	71286	71394	64472
315 807	**GE**	H	*GE*	IL	64473	71287	71395	64474
315 808	**GE**	H	*GE*	IL	64475	71288	71396	64476
315 809	**GE**	H	*GE*	IL	64477	71289	71397	64478
315 810	**GE**	H	*GE*	IL	64479	71290	71398	64480
315 811	**GE**	H	*GE*	IL	64481	71291	71399	64482
315 812	**GE**	H	*GE*	IL	64483	71292	71400	64484
315 813	**GE**	H	*GE*	IL	64485	71293	71401	64486
315 814	**GE**	H	*GE*	IL	64487	71294	71402	64488
315 815	**GE**	H	*GE*	IL	64489	71295	71403	64490
315 816	**GE**	H	*GE*	IL	64491	71296	71404	64492
315 817	**GE**	H	*GE*	IL	64493	71297	71405	64494
315 818	**GE**	H	*GE*	IL	64495	71298	71406	64496
315 819	**GE**	H	*GE*	IL	64497	71299	71407	64498
315 820	**GE**	H	*GE*	IL	64499	71300	71408	64500
315 821	**GE**	H	*GE*	IL	64501	71301	71409	64502
315 822	**GE**	H	*GE*	IL	64503	71302	71410	64504
315 823	**GE**	H	*GE*	IL	64505	71303	71411	64506
315 824	**GE**	H	*GE*	IL	64507	71304	71412	64508
315 825	**GE**	H	*GE*	IL	64509	71305	71413	64510
315 826	**GE**	H	*GE*	IL	64511	71306	71414	64512
315 827	**GE**	H	*GE*	IL	64513	71307	71415	64514
315 828	**GE**	H	*GE*	IL	64515	71308	71416	64516
315 829	**GE**	H	*GE*	IL	64517	71309	71417	64518
315 830	**GE**	H	*GE*	IL	64519	71310	71418	64520

315 831	**GE**	H	*GE*	IL	64521	71311	71419	64522
315 832	**GE**	H	*GE*	IL	64523	71312	71420	64524
315 833	**GE**	H	*GE*	IL	64525	71313	71421	64526
315 834	**GE**	H	*GE*	IL	64527	71314	71422	64528
315 835	**GE**	H	*GE*	IL	64529	71315	71423	64530
315 836	**GE**	H	*GE*	IL	64531	71316	71424	64532
315 837	**GE**	H	*GE*	IL	64533	71317	71425	64534
315 838	**GE**	H	*GE*	IL	64535	71318	71426	64536
315 839	**GE**	H	*GE*	IL	64537	71319	71427	64538
315 840	**GE**	H	*GE*	IL	64539	71320	71428	64540
315 841	**GE**	H	*GE*	IL	64541	71321	71429	64542

Units 315 842–315 861. Brush traction motors.

315 842	**GE**	H	*GE*	IL	64543	71322	71430	64544
315 843	**GE**	H	*GE*	IL	64545	71323	71431	64546
315 844	**AL**	H	*WN*	HE	64547	71324	71432	64548
315 845	**U**	H	*WN*	HE	64549	71325	71433	64550
315 846	**U**	H	*WN*	HE	64551	71326	71434	64552
315 847	**U**	H	*WN*	HE	64553	71327	71435	64554
315 848	**N**	H	*WN*	HE	64555	71328	71436	64556
315 849	**U**	H	*WN*	HE	64557	71329	71437	64558
315 850	**N**	H	*WN*	HE	64559	71330	71438	64560
315 851	**N**	H	*WN*	HE	64561	71331	71439	64562
315 852	**N**	H	*WN*	HE	64563	71332	71440	64564
315 853	**N**	H	*WN*	HE	64565	71333	71441	64566
315 854	**N**	H	*WN*	HE	64567	71334	71442	64568
315 855	**N**	H	*WN*	HE	64569	71335	71443	64570
315 856	**N**	H	*WN*	HE	64571	71336	71444	64572
315 857	**N**	H	*WN*	HE	64573	71337	71445	64574
315 858	**N**	H	*WN*	HE	64579	71338	71446	64580
315 859	**N**	H	*WN*	HE	64577	71339	71447	64578
315 860	**N**	H	*WN*	HE	64575	71340	71448	64576
315 861	**N**	H	*WN*	HE	64581	71341	71449	64582

CLASS 317 4-Car Unit

Various formations, see below. Gangwayed throughout. Disc brakes.
Traction Motors: Four GEC G315BZ of 247.5 kW each.
Dimensions: 20.13 (DTS & DTC) or 20.18 (ATC, ATS & PMS) x 2.82 x 3.58 m.
Maximum Speed: 100 mph. **Doors**: Power operated sliding.
Couplers: Tightlock. **Bogies**: BREL BP20/BT13.
Multiple Working: Classes 313–323.

Class 317/1 († 317/3). Pressure heating & ventilation. DTS(A)–PMS–ATC († ATS)–DTS(B).

62661–62708. PMS. Dia. EC208. Lot No. 30958 BREL York 1981–82. –/79. 49.8 t.
71577–71624. ATC († ATS). Dia. EH307 († EH242). Lot No. 30957 BREL Derby 1981–82. 22/46 († –/68) 2T. 28.8 t.
77000–77047. DTS(A). Dia. EE216. Lot No. 30955 BREL York 1981–82. –/74. 29.4 t.
77048–77095. DTS(B). Dia. EE235 (*EE232). Lot No. 30956 York 1981–82. –/70. (* –/71). 29.3 t.

317 301	†	**LS**	A	*LS*	EM	77024	62661	71577	77048
317 302	†	**LS**	A	*LS*	EM	77001	62662	71578	77049
317 303	†	**LS**	A	*LS*	EM	77002	62663	71579	77050
317 304	†	**LS**	A	*LS*	EM	77003	62664	71580	77051
317 305	†	**LS**	A	*LS*	EM	77004	62665	71581	77052
317 306	†	**LS**	A	*LS*	EM	77005	62666	71582	77053
317 307	†	**LS**	A	*LS*	EM	77006	62667	71583	77054
317 309		**N**	A	*WN*	HE	77008	62669	71585	77056
317 310		**N**	A	*WN*	HE	77009	62670	71586	77057
317 311	†	**LS**	A	*LS*	EM	77010	62697	71587	77058
317 312	†	**LS**	A	*LS*	EM	77011	62672	71588	77059
317 313	†	**LS**	A	*LS*	EM	77012	62673	71589	77060
317 314		**N**	A	*WN*	HE	77013	62674	71590	77061
317 315		**N**	A	*WN*	HE	77014	62675	71591	77062
317 316		**N**	A	*WN*	HE	77015	62676	71592	77063
317 317		**N**	A	*TR*	HE	77016	62677	71593	77064
317 318		**N**	A	*WN*	HE	77017	62678	71594	77065
317 319	†	**LS**	A	*LS*	EM	77018	62679	71595	77066
317 320		**N**	A	*WN*	HE	77019	62680	71596	77067
317 321		**N**	A	*WN*	HE	77020	62681	71597	77068
317 324		**N**	A	*TR*	HE	77023	62684	71600	77071
317 325		**N**	A	*WN*	HE	77000	62685	71601	77072
317 326		**N**	A	*TR*	HE	77025	62686	71602	77073
317 327		**N**	A	*WN*	HE	77026	62687	71603	77074
317 328		**N**	A	*WN*	HE	77027	62688	71604	77075
317 330		**N**	A	*WN*	HE	77043	62704	71606	77077
317 331		**N**	A	*WN*	HE	77030	62691	71607	77078
317 332	†	**LS**	A	*LS*	EM	77031	62692	71608	77079
317 333		**N**	A	*WN*	HE	77032	62693	71609	77080
317 334		**N**	A	*WN*	HE	77033	62694	71610	77081
317 335		**N**	A	*WN*	HE	77034	62695	71611	77082
317 336		**N**	A	*WN*	HE	77035	62696	71612	77083

317 337	*	**N**	A	*WN*	HE	77036	62671	71613	77084
317 338	*	**N**	A	*WN*	HE	77037	62698	71614	77085
317 339	*	**N**	A	*WN*	HE	77038	62699	71615	77086
317 340	*	**N**	A	*WN*	HE	77039	62700	71616	77087
317 341	*	**N**	A	*WN*	HE	77040	62701	71617	77088
317 342	*	**N**	A	*WN*	HE	77041	62702	71618	77089
317 343	*	**N**	A	*WN*	HE	77042	62703	71619	77090
317 344	*	**N**	A	*WN*	HE	77029	62690	71620	77091
317 345	*	**N**	A	*WN*	HE	77044	62705	71621	77092
317 346	*	**N**	A	*WN*	HE	77045	62706	71622	77093
317 347	*	**N**	A	*WN*	HE	77046	62707	71623	77094
317 348	*	**N**	A	*WN*	HE	77047	62708	71624	77095
317 392	†	**LS**	A	*LS*	EM	77021	62682	71598	77069
317 393	†	**LS**	A	*TR*	HE	77022	62683	71599	77070

Class 317/6. Convection heating. DTS–PMS–ATS–DTC.

62846–62865. PMS. Dia. EC222. Lot No. 30996 BREL York 1985–86. Refurbished Railcare Wolverton 1998–99. –/70. 50.1 t.
62886–62889. PMS. Dia. EC222. Lot No. 31009 BREL York 1987. Refurbished Railcare Wolverton 1998–99. –/70. 50.1 t.
71734–71753. ATS. Dia. EH247. Lot No. 30997 BREL York 1985–86. Refurbished Railcare Wolverton 1998–99. –/62 2T. 28.8 t.
71762–71765. ATS. Dia. EH247. Lot No. 31010 BREL York 1987. Refurbished Railcare Wolverton 1998–99. –/62 2T. 28.8 t.
77200–77219. DTS. Dia. EE247. Lot No. 30994 BREL York 1985–86. Refurbished Railcare Wolverton 1998–99. –/64. 29.3 t.
77280–77283. DTS. Dia. EE247. Lot No. 31007 BREL York 1987. Refurbished Railcare Wolverton 1998–99. –/64. 29.3 t.
77220–77239. DTC. Dia. EE375. Lot No. 30995 BREL York 1985–86. Refurbished Railcare Wolverton 1998–99. 24/48. 29.3 t.
77284–77287. DTC. Dia. EE375. Lot No. 31008 BREL York 1987. Refurbished Railcare Wolverton 1998–99. 24/48. 29.3 t.

Number	*Former No.*								
317 649	317 349	**WN**	A	*WN*	HE	77200	62846	71734	77220
317 650	317 350	**WN**	A	*WN*	HE	77201	62847	71735	77221
317 651	317 351	**WN**	A	*WN*	HE	77202	62848	71736	77222
317 652	317 352	**WN**	A	*WN*	HE	77203	62849	71739	77223
317 653	317 353	**WN**	A	*WN*	HE	77204	62850	71738	77224
317 654	317 354	**WN**	A	*WN*	HE	77205	62851	71737	77225
317 655	317 355	**WN**	A	*WN*	HE	77206	62852	71740	77226
317 656	317 356	**WN**	A	*WN*	HE	77207	62853	71742	77227
317 657	317 357	**WN**	A	*WN*	HE	77208	62854	71741	77228
317 658	317 358	**WN**	A	*WN*	HE	77209	62855	71743	77229
317 659	317 359	**WN**	A	*WN*	HE	77210	62856	71744	77230
317 660	317 360	**WN**	A	*WN*	HE	77211	62857	71745	77231
317 661	317 361	**WN**	A	*WN*	HE	77212	62858	71746	77232
317 662	317 362	**WN**	A	*WN*	HE	77213	62859	71747	77233
317 663	317 363	**WN**	A	*WN*	HE	77214	62860	71748	77234
317 664	317 364	**WN**	A	*WN*	HE	77215	62861	71749	77235
317 665	317 365	**WN**	A	*WN*	HE	77216	62862	71750	77236

317 666	317 366	**WN**	A	*WN*	HE	77217	62863	71752	77237
317 667	317 367	**WN**	A	*WN*	HE	77218	62864	71751	77238
317 668	317 368	**WN**	A	*WN*	HE	77219	62865	71753	77239
317 669	317 369	**WN**	A	*WN*	HE	77280	62886	71762	77284
317 670	317 370	**WN**	A	*WN*	HE	77281	62887	71763	77285
317 671	317 371	**WN**	A	*WN*	HE	77282	62888	71764	77286
317 672	317 372	**WN**	A	*WN*	HE	77283	62889	71765	77287

Class 317/7. Dedicated units to be converted for services to/from Stansted Airport. Pressure heating & ventilation. Final details awaited.

62661–62708. PMS. Dia. EC2??. Lot No. 30958 BREL York 1981-82. Refurbished Railcare Wolverton 1999–2000. –/64.
71577–71624. TS. Dia. EH2??. Lot No. 30957 BREL Derby 1981-82. Refurbished Railcare Wolverton 1999–2000. –/46. 1T 1TD.
77000–77047. DTS. Dia. EE2??. Lot No. 30955 BREL York 1981-82. Refurbished Railcare Wolverton 1999–2000. –/54.
77048–77095. DTC. Dia. EE3??. Lot No. 30956 York 1981–82. Refurbished Railcare Wolverton 1999–2000. 23/16.

Number	*Former No.*							
317 701	317 308	A	*WN*	HE	77007	62668	71584	77055
317 702	317 329	A	*WN*	HE	77028	62689	71605	77076
317 703	317 3							
317 704	317 3							
317 705	317 3							
317 706	317 3							
317 707	317 3							
317 708	317 3							
317 709	317 3							

CLASS 318 3-Car Unit

DTS(A)–PMS–DTS(B). Gangwayed throughout. Disc brakes.
Traction Motors: Four Brush TM 2141 of 268 kW each.
Dimensions: 20.13 (DTS) or 20.18 (PMS) x 2.82 x 3.77 m.
Maximum Speed: 90 mph. **Doors**: Power operated sliding.
Couplers: Tightlock. **Bogies**: BREL BP20/BT13.
Multiple Working: Classes 313–323.

62866–62885. PMS. Dia. EC207. Lot No. 30998 BREL York 1985–86. –/79. 50.9 t.
62890. PMS. Dia. EC207. Lot No. 31019 BREL York 1987. –/79. 50.9 t.
77240–77259. DTS(B). Dia. EE227. Lot No. 30999 BREL York 1985–86. –/66 1T. 30.0 t.
77260–77279. DTS(A). Dia. EE228. Lot No. 31000 BREL York 1985–86. –/71. 26.6 t.
77288. DTS(B). Dia. EE227. Lot No. 31020 BREL York 1987. –/66 1T. 30.0 t.
77289. DTS(A). Dia. EE228. Lot No. 31021 BREL York 1987. –/71. 26.6 t.

318 250	**S**	H	*SR*	GW	77260	62866	77240
318 251	**CC**	H	*SR*	GW	77261	62867	77241
318 252	**CC**	H	*SR*	GW	77262	62868	77242
318 253	**CC**	H	*SR*	GW	77263	62869	77243

318 254	**CC**	H	*SR*	GW	77264	62870	77244
318 255	**S**	H	*SR*	GW	77265	62871	77245
318 256	**CC**	H	*SR*	GW	77266	62872	77246
318 257	**CC**	H	*SR*	GW	77267	62873	77247
318 258	**CC**	H	*SR*	GW	77268	62874	77248
318 259	**CC**	H	*SR*	GW	77269	62875	77249
318 260	**S**	H	*SR*	GW	77270	62876	77250
318 261	**S**	H	*SR*	GW	77271	62877	77251
318 262	**CC**	H	*SR*	GW	77272	62878	77252
318 263	**CC**	H	*SR*	GW	77273	62879	77253
318 264	**CC**	H	*SR*	GW	77274	62880	77254
318 265	**S**	H	*SR*	GW	77275	62881	77255
318 266	**CC**	H	*SR*	GW	77276	62882	77256
318 267	**S**	H	*SR*	GW	77277	62883	77257
318 268	**S**	H	*SR*	GW	77278	62884	77258
318 269	**CC**	H	*SR*	GW	77279	62885	77259
318 270	**CC**	H	*SR*	GW	77289	62890	77288

Names (carried on PMS):

318 259 Citizens' Network | 318 266 STRATHCLYDER

CLASS 319 4-Car Unit

Various formations, see below. Gangwayed within unit. End doors. Disc brakes.
Supply System: 25 kV 50 Hz a.c. overhead and/or 750 V d.c. third rail.
Traction Motors: Four GEC G315BZ of 247.5 kW each.
Dimensions: 20.13 (DTC & DTS) or 20.18 (ATS & PMS) x 2.82 x 3.77 m.
Maximum Speed: 100 mph. **Doors**: Power operated sliding.
Couplers: Tightlock. **Bogies**: BREL P7-4/T3-7.
Multiple Working: Classes 313–323.

Class 319/0. DTS(A)–PMS–ATS–DTS(B). 25 kV 50 Hz a.c. overhead or 750 V d.c. third rail supply.

62891–62903. PMS. Dia. EC209. Lot No. 31023 BREL York 1987–8. –/77 2T. 51.0 t.
71772–71784. ATS. Dia. EH234. Lot No. 31024 BREL York 1987–8. –/77 2T. 31.0 t.
77290–77314 (Even numbers). DTS(B). Dia. EE234. Lot No. 31025 BREL York 1987–8. –/78. 30.0 t.
77291–77315 (Odd numbers). DTS(A). Dia. EE233. Lot No. 31022 BREL York 1987–8. –/82. 30.0 t.

319 001	**CX**	P	*SC*	SU	77291	62891	71772	77290
319 002	**CX**	P	*TR*	SU	77293	62892	71773	77292
319 003	**CX**	P	*SC*	SU	77295	62893	71774	77294
319 004	**CX**	P	*SC*	SU	77297	62894	71775	77296
319 005	**CX**	P	*SC*	SU	77299	62895	71776	77298
319 006	**CX**	P	*SC*	SU	77301	62896	71777	77300
319 007	**CX**	P	*TR*	SU	77303	62897	71778	77302
319 008	**CX**	P	*SC*	SU	77305	62898	71779	77304
319 009	**CX**	P	*SC*	SU	77307	62899	71780	77306

319 010	**CX**	P	*SC*	SU	77309	62900	71781	77308
319 011	**CX**	P	*SC*	SU	77311	62901	71782	77310
319 012	**CX**	P	*TR*	SU	77313	62902	71783	77312
319 013	**CX**	P	*SC*	SU	77315	62903	71784	77314

Names (carried on ATS):

319 005 Partnership For Progress
319 008 Cheriton
319 009 Coquelles
319 011 John Ruskin College
319 013 The Surrey Hills

Class 319/2. DTS–PMB–ATS–DTC. 750 V d.c. third rail with provision for 25 kV a.c. 50 Hz overhead supply. 'Express' configuration units operated by Connex South Central on the London Victoria–Brighton route.

62904–62910. PMB. Dia. EN262. Lot No. 31023 BREL York 1987–8. Refurbished Railcare Wolverton 1996. –/60 2T. 51.0 t.
71785–71791. ATS. Dia. EH212. Lot No. 31024 BREL York 1987–8. Refurbished Railcare Wolverton 1996. –/52 1T 1TD. 31.0 t.
77316–77328 (Even numbers). DTC. Dia. EE374. Lot No. 31025 BREL York 1987–8. Refurbished Railcare Wolverton 1996. 18/36. 29.0 t.
77317–77329 (Odd numbers). DTS. Dia. EE244. Lot No. 31022 BREL York 1987–8. Refurbished Railcare Wolverton 1996. –/64. 29.7 t.

319 214	**CX**	P	*SC*	SU	77317	62904	71785	77316
319 215	**CX**	P	*SC*	SU	77319	62905	71786	77318
319 216	**CX**	P	*SC*	SU	77321	62906	71787	77320
319 217	**CX**	P	*SC*	SU	77323	62907	71788	77322
319 218	**CX**	P	*SC*	SU	77325	62908	71789	77324
319 219	**CX**	P	*SC*	SU	77327	62909	71790	77326
319 220	**CX**	P	*SC*	SU	77329	62910	71791	77328

Names (carried on ATS):

319 215 London
319 217 Brighton
319 218 Croydon

Class 319/3. DTS(A)–PMS–ATS–DTS(B). 25 kV 50 Hz a.c. overhead or 750 V d.c. third rail supply. Operated by Thameslink Rail on the 'City Metro' Luton–Sutton route.

63043–63062, 63098–63098. PMS. Dia. EC214. Lot No. 31064 BREL York 1990. –/79. 50.3 t.
71929–71948, 71979–71984. ATS. Dia. EH238. Lot No. 31065 BREL York 1990. –/74 2T. 33.1 t.
77458–77496, 77974–77984 (Even numbers). DTS(B). Dia. EE240. Lot No. 31066 BREL York 1990. –/78. 29.7 t.
77459–77497, 77973–77983 (Odd numbers). DTS(A). Dia. EE240. Lot No. 31063 BREL York 1990. –/70. 29.0 t.

Number	*Former No.*								
319 361	319 161	**TR**	P	*TR*	SU	77459	63043	71929	77458
319 362	319 162	**TR**	P	*TR*	SU	77461	63044	71930	77460
319 363	319 163	**TR**	P	*TR*	SU	77463	63045	71931	77462
319 364	319 164	**TR**	P	*TR*	SU	77465	63046	71932	77464
319 365	319 165	**TR**	P	*TR*	SU	77467	63047	71933	77466

319 366	319 166	**TR**	P	*TR*	SU	77469	63048	71934	77468
319 367	319 167	**TR**	P	*TR*	SU	77471	63049	71935	77470
319 368	319 168	**TR**	P	*TR*	SU	77473	63050	71936	77472
319 369	319 169	**TR**	P	*TR*	SU	77475	63051	71937	77474
319 370	319 170	**TR**	P	*TR*	SU	77477	63052	71938	77476
319 371	319 171	**TR**	P	*TR*	SU	77479	63053	71939	77478
319 372	319 172	**TR**	P	*TR*	SU	77481	63054	71940	77480
319 373	319 173	**TR**	P	*TR*	SU	77483	63055	71941	77482
319 374	319 174	**TR**	P	*TR*	SU	77485	63056	71942	77484
319 375	319 175	**TR**	P	*TR*	SU	77487	63057	71943	77486
319 376	319 176	**TR**	P	*TR*	SU	77489	63058	71944	77488
319 377	319 177	**TR**	P	*TR*	SU	77491	63059	71945	77490
319 378	319 178	**TR**	P	*TR*	SU	77493	63060	71946	77492
319 379	319 179	**TR**	P	*TR*	SU	77495	63061	71947	77494
319 380	319 180	**TR**	P	*TR*	SU	77497	63062	71948	77496
319 381	319 181	**TR**	P	*TR*	SU	77973	63093	71979	77974
319 382	319 182	**TR**	P	*TR*	SU	77975	63094	71980	77976
319 383	319 183	**TR**	P	*TR*	SU	77977	63095	71981	77978
319 384	319 184	**TR**	P	*TR*	SU	77979	63096	71982	77980
319 385	319 185	**TR**	P	*TR*	SU	77981	63097	71983	77982
319 386	319 186	**TR**	P	*TR*	SU	77983	63098	71984	77984

Class 319/4. DTC–PMS–ATS–DTS. 25 kV 50 Hz a.c. overhead or 750 V d.c. third rail supply. Operated by Thameslink Rail on the 'City Flyer' Bedford–Gatwick Airport–Brighton route.

62911–62936. PMS. Dia. EC209. Lot No. 31023 BREL York 1987–88. –/77 2T. 51.0 t.
62961–62974. PMS. Dia. EC209. Lot No. 31039 BREL York 1988. –/77 2T. 51.0 t.
71792–71817. ATS. Dia. EH234. Lot No. 31024 BREL York 1987–88. –/77 2T. 34.0 t.
71866–71879. ATS. Dia. EH234. Lot No. 31040 BREL York 1988. –/77 2T. 34.0 t.
77330–77380 (Even numbers). DTS. Dia. EE234. Lot No. 31025 BREL York 1987-88. –/78. 30.0 t.
77331–77381 (Odd numbers). DTC. Dia. EE314. Lot No. 31022 BREL York 1987–88. 12/54. 30.0 t.
77430–77456 (Even numbers). DTS. Dia. EE234. Lot No. 31041 BREL York 1988. –/74. 30.0 t.
77431–77457 (Odd numbers). DTC. Dia. EE314. Lot No. 31038 BREL York 1988. 12/54. 30.0 t.

319 421	**TR**	P	*TR*	SU	77331	62911	71792	77330
319 422	**TR**	P	*TR*	SU	77333	62912	71793	77332
319 423	**TR**	P	*TR*	SU	77335	62913	71794	77334
319 424	**TR**	P	*TR*	SU	77337	62914	71795	77336
319 425	**TR**	P	*TR*	SU	77339	62915	71796	77338
319 426	**TR**	P	*TR*	SU	77341	62916	71797	77340
319 427	**TR**	P	*TR*	SU	77343	62917	71798	77342
319 428	**TR**	P	*TR*	SU	77345	62918	71799	77344
319 429	**TR**	P	*TR*	SU	77347	62919	71800	77346
319 430	**TR**	P	*TR*	SU	77349	62920	71801	77348
319 431	**TR**	P	*TR*	SU	77351	62921	71802	77350

319 432	**TR**	P	*TR*	SU	77353	62922	71803	77352
319 433	**TR**	P	*TR*	SU	77355	62923	71804	77354
319 434	**TR**	P	*TR*	SU	77357	62924	71805	77356
319 435	**TR**	P	*TR*	SU	77359	62925	71806	77358
319 436	**TR**	P	*TR*	SU	77361	62926	71807	77360
319 437	**TR**	P	*TR*	SU	77363	62927	71808	77362
319 438	**TR**	P	*TR*	SU	77365	62928	71809	77364
319 439	**TR**	P	*TR*	SU	77367	62929	71810	77366
319 440	**TR**	P	*TR*	SU	77369	62930	71811	77368
319 441	**TR**	P	*TR*	SU	77371	62931	71812	77370
319 442	**TR**	P	*TR*	SU	77373	62932	71813	77372
319 443	**TR**	P	*TR*	SU	77375	62933	71814	77374
319 444	**TR**	P	*TR*	SU	77377	62934	71815	77376
319 445	**TR**	P	*TR*	SU	77379	62935	71816	77378
319 446	**TR**	P	*TR*	SU	77381	62936	71817	77380
319 447	**TR**	P	*TR*	SU	77431	62961	71866	77430
319 448	**TR**	P	*TR*	SU	77433	62962	71867	77432
319 449	**TR**	P	*TR*	SU	77435	62963	71868	77434
319 450	**TR**	P	*TR*	SU	77437	62964	71869	77436
319 451	**TR**	P	*TR*	SU	77439	62965	71870	77438
319 452	**TR**	P	*TR*	SU	77441	62966	71871	77440
319 453	**TR**	P	*TR*	SU	77443	62967	71872	77442
319 454	**TR**	P	*TR*	SU	77445	62968	71873	77444
319 455	**TR**	P	*TR*	SU	77447	62969	71874	77446
319 456	**TR**	P	*TR*	SU	77449	62970	71875	77448
319 457	**TR**	P	*TR*	SU	77451	62971	71876	77450
319 458	**TR**	P	*TR*	SU	77453	62972	71877	77452
319 459	**TR**	P	*TR*	SU	77455	62973	71878	77454
319 460	**TR**	P	*TR*	SU	77457	62974	71879	77456

CLASS 320 3-Car Unit

DTS(A)–PMS–DTS(B). Gangwayed within unit. Disc brakes.
Traction Motors: Four Brush TM2141B of 268 kW each.
Dimensions: 19.95 (DTS) or 19.92 (PMS) x 2.82 x 3.78 m.
Maximum Speed: 75 mph. **Doors**: Power operated sliding.
Couplers: Tightlock. **Bogies**: BREL P7-4/T3-7.
Multiple Working: Classes 313–323.

63021–63042. PMS. Dia. EC212. Lot No. 31062 BREL York 1990. –/77. 52.1 t.
77899–77920. DTS(A). Dia. EE238. Lot No. 31060 BREL York 1990. –/77. 30.7 t.
77921–77942. DTS(B). Dia. EE239. Lot No. 31061 BREL York 1990. –/76 31.7 t.

320 301	**S**	H	*SR*	GW	77899	63021	77921
320 302	**S**	H	*SR*	GW	77900	63022	77922
320 303	**S**	H	*SR*	GW	77901	63023	77923
320 304	**S**	H	*SR*	GW	77902	63024	77924
320 305	**S**	H	*SR*	GW	77903	63025	77925
320 306	**CC**	H	*SR*	GW	77904	63026	77926
320 307	**CC**	H	*SR*	GW	77905	63027	77927
320 308	**CC**	H	*SR*	GW	77906	63028	77928

320 309	**CC**	H	*SR*	GW	77907	63029	77929
320 310	**CC**	H	*SR*	GW	77908	63030	77930
320 311	**CC**	H	*SR*	GW	77909	63031	77931
320 312	**CC**	H	*SR*	GW	77910	63032	77932
320 313	**CC**	H	*SR*	GW	77911	63033	77933
320 314	**CC**	H	*SR*	GW	77912	63034	77934
320 315	**CC**	H	*SR*	GW	77913	63035	77935
320 316	**CC**	H	*SR*	GW	77914	63036	77936
320 317	**CC**	H	*SR*	GW	77915	63037	77937
320 318	**CC**	H	*SR*	GW	77916	63038	77938
320 319	**CC**	H	*SR*	GW	77917	63039	77939
320 320	**CC**	H	*SR*	GW	77918	63040	77940
320 321	**CC**	H	*SR*	GW	77919	63041	77941
320 322	**CC**	H	*SR*	GW	77920	63042	77942

Names (carried on PMS):

320 305 GLASGOW SCHOOL OF ART 1844–150–1994
320 306 MODEL RAIL SCOTLAND
320 309 Radio Clyde 25th Anniversary
320 311 The Royal College of Physicians and Surgeons of Glasgow
320 321 The Rt. Hon. John Smith, QC, MP
320 322 FESTIVE GLASGOW ORCHID

CLASS 321 4-Car Unit

Various formations, see below. Gangwayed within unit. Disc brakes.
Traction Motors: Four Brush TM2141B of 268 kW each.
Dimensions: 19.95 (DTC & DTS) or 19.92 (ATS & PMS) x 2.82 x 3.78 m.
Maximum Speed: 100 mph.
Doors: Power operated sliding.
Couplers: Tightlock.
Bogies: BREL P7-4/T3-7.
Multiple Working: Classes 313–323.

Class 321/3. DTC–PMS–ATS–DTS. Small first class area.

62975–63020, 63105–63124. PMS. Dia. EC210. Lot No. 31054 BREL York 1988–90. –/79. 51.5 t.
71880–71925, 71991–72010. ATS. Dia. EH235. Lot No. 31055 BREL York 1988–90. –/74 2T. 28.0 t.
77853–77898, 78280–78299. DTS. Dia. EE236. Lot No. 31056 BREL York 1988–90. –/78. 29.1 t.
78049–78094, 78131–78150. DTC. Dia. EE308. Lot No. 31053 BREL York 1988–90. 12/56. 29.3 t.

321 301	**GE**	H	*GE*	IL	78049	62975	71880	77853
321 302	**GE**	H	*GE*	IL	78050	62976	71881	77854
321 303	**GE**	H	*GE*	IL	78051	62977	71882	77855
321 304	**GE**	H	*GE*	IL	78052	62978	71883	77856
321 305	**GE**	H	*GE*	IL	78053	62979	71884	77857
321 306	**GE**	H	*GE*	IL	78054	62980	71885	77858
321 307	**GE**	H	*GE*	IL	78055	62981	71886	77859
321 308	**GE**	H	*GE*	IL	78056	62982	71887	77860
321 309	**GE**	H	*GE*	IL	78057	62983	71888	77861

321 310	**GE**	H	*GE*	IL	78058	62984	71889	77862
321 311	**GE**	H	*GE*	IL	78059	62985	71890	77863
321 312	**GE**	H	*GE*	IL	78060	62986	71891	77864
321 313	**GE**	H	*GE*	IL	78061	62987	71892	77865
321 314	**GE**	H	*GE*	IL	78062	62988	71893	77866
321 315	**GE**	H	*GE*	IL	78063	62989	71894	77867
321 316	**GE**	H	*GE*	IL	78064	62990	71895	77868
321 317	**GE**	H	*GE*	IL	78065	62991	71896	77869
321 318	**GE**	H	*GE*	IL	78066	62992	71897	77870
321 319	**GE**	H	*GE*	IL	78067	62993	71898	77871
321 320	**GE**	H	*GE*	IL	78068	62994	71899	77872
321 321	**GE**	H	*GE*	IL	78069	62995	71900	77873
321 322	**GE**	H	*GE*	IL	78070	62996	71901	77874
321 323	**GE**	H	*GE*	IL	78071	62997	71902	77875
321 324	**GE**	H	*GE*	IL	78072	62998	71903	77876
321 325	**GE**	H	*GE*	IL	78073	62999	71904	77877
321 326	**GE**	H	*GE*	IL	78074	63000	71905	77878
321 327	**GE**	H	*GE*	IL	78075	63001	71906	77879
321 328	**GE**	H	*GE*	IL	78076	63002	71907	77880
321 329	**GE**	H	*GE*	IL	78077	63003	71908	77881
321 330	**GE**	H	*GE*	IL	78078	63004	71909	77882
321 331	**GE**	H	*GE*	IL	78079	63005	71910	77883
321 332	**GE**	H	*GE*	IL	78080	63006	71911	77884
321 333	**GE**	H	*GE*	IL	78081	63007	71912	77885
321 334	**GE**	H	*GE*	IL	78082	63008	71913	77886
321 335	**GE**	H	*GE*	IL	78083	63009	71914	77887
321 336	**GE**	H	*GE*	IL	78084	63010	71915	77888
321 337	**GE**	H	*GE*	IL	78085	63011	71916	77889
321 338	**GE**	H	*GE*	IL	78086	63012	71917	77890
321 339	**GE**	H	*GE*	IL	78087	63013	71918	77891
321 340	**GE**	H	*GE*	IL	78088	63014	71919	77892
321 341	**GE**	H	*GE*	IL	78089	63015	71920	77893
321 342	**GE**	H	*GE*	IL	78090	63016	71921	77894
321 343	**GE**	H	*GE*	IL	78091	63017	71922	77895
321 344	**GE**	H	*GE*	IL	78092	63018	71923	77896
321 345	**GE**	H	*GE*	IL	78093	63019	71924	77897
321 346	**GE**	H	*GE*	IL	78094	63020	71925	77898
321 347	**GE**	H	*GE*	IL	78131	63105	71991	78280
321 348	**GE**	H	*GE*	IL	78132	63106	71992	78281
321 349	**GE**	H	*GE*	IL	78133	63107	71993	78282
321 350	**GE**	H	*GE*	IL	78134	63108	71994	78283
321 351	**GE**	H	*GE*	IL	78135	63109	71995	78284
321 352	**GE**	H	*GE*	IL	78136	63110	71996	78285
321 353	**GE**	H	*GE*	IL	78137	63111	71997	78286
321 354	**GE**	H	*GE*	IL	78138	63112	71998	78287
321 355	**GE**	H	*GE*	IL	78139	63113	71999	78288
321 356	**GE**	H	*GE*	IL	78140	63114	72000	78289
321 357	**GE**	H	*GE*	IL	78141	63115	72001	78290
321 358	**GE**	H	*GE*	IL	78142	63116	72002	78291
321 359	**GE**	H	*GE*	IL	78143	63117	72003	78292
321 360	**GE**	H	*GE*	IL	78144	63118	72004	78293

321 361	**GE**	H	*GE*	IL	78145	63119	72005	78294
321 362	**GE**	H	*GE*	IL	78146	63120	72006	78295
321 363	**GE**	H	*GE*	IL	78147	63121	72007	78296
321 364	**GE**	H	*GE*	IL	78148	63122	72008	78297
321 365	**GE**	H	*GE*	IL	78149	63123	72009	78298
321 366	**GE**	H	*GE*	IL	78150	63124	72010	78299

Names (carried on ATS):

321 312 Southend-on-Sea
321 334 Amsterdam
321 336 GEOFFREY FREEMAN ALLEN
321 351 GURKHA

Class 321/4. DTC–PMS–ATS–DTS. Large first class area. All First Great Eastern operated DTC have 12 first class seats declassfied.

Advertising Livery:
- 321 428 'Birmingham Daytripper Ticket'.

63063–63092, 63099–63104, 63125–63136. PMS. Dia. EC210. Lot No. 31068 BREL York 1989–90. –/79. 51.5 t.
63082″. PMS. Dia. EC210. Adtranz Crewe 1998. –/79. 51.5 t.
71949–71978, 71985–71990, 72011–72022. ATS. Dia. EH235. Lot No. 31069 BREL York 1989–90. –/74 2T. 28.0 t.
71966″. ATS. Dia. EH235. Adtranz Crewe 1998. –/74 2T. 28.0 t.
77943–77972, 78274–78279, 78300–78311. DTS. Dia. EE236. Lot No. 31070 BREL York 1989–90. –/78. 29.1 t.
77960″. DTS. Dia. EE236. Adtranz Crewe 1998. –/78. 29.1 t.
78095–78130/151–78162. DTC. Dia. EE309. Lot No. 31067 BREL York 1989–90. 28/40. 29.3 t.
78114″. DTC. Dia. EE309. Adtranz Crewe 1998. 28/40. 29.3 t.

321 401	**N**	H	*SL*	BY	78095	63063	71949	77943
321 402	**N**	H	*SL*	BY	78096	63064	71950	77944
321 403	**N**	H	*SL*	BY	78097	63065	71951	77945
321 404	**N**	H	*SL*	BY	78098	63066	71952	77946
321 405	**N**	H	*SL*	BY	78099	63067	71953	77947
321 406	**N**	H	*SL*	BY	78100	63068	71954	77948
321 407	**N**	H	*SL*	BY	78101	63069	71955	77949
321 408	**N**	H	*SL*	BY	78102	63070	71956	77950
321 409	**N**	H	*SL*	BY	78103	63071	71957	77951
321 410	**N**	H	*SL*	BY	78104	63072	71958	77952
321 411	**N**	H	*SL*	BY	78105	63073	71959	77953
321 412	**N**	H	*SL*	BY	78106	63074	71960	77954
321 413	**N**	H	*SL*	BY	78107	63075	71961	77955
321 414	**N**	H	*SL*	BY	78108	63076	71962	77956
321 415	**N**	H	*SL*	BY	78109	63077	71963	77957
321 416	**N**	H	*SL*	BY	78110	63078	71964	77958
321 417	**N**	H	*SL*	BY	78111	63079	71965	77959
321 418	**N**	H	*SL*	BY	78112	63080	71968	77962
321 419	**N**	H	*SL*	BY	78113	63081	71967	77961
321 420	**SL**	H	*SL*	BY	78114″	63082″	71966″	77960″
321 421	**N**	H	*SL*	BY	78115	63083	71969	77963

321 422	**SL**	H	*SL*	BY	78116	63084	71970	77964
321 423	**N**	H	*SL*	BY	78117	63085	71971	77965
321 424	**N**	H	*SL*	BY	78118	63086	71972	77966
321 425	**N**	H	*SL*	BY	78119	63087	71973	77967
321 426	**N**	H	*SL*	BY	78120	63088	71974	77968
321 427	**N**	H	*SL*	BY	78121	63089	71975	77969
321 428	**AL**	H	*SL*	BY	78122	63090	71976	77970
321 429	**SL**	H	*SL*	BY	78123	63091	71977	77971
321 430	**SL**	H	*SL*	BY	78124	63092	71978	77972
321 431	**SL**	H	*SL*	BY	78151	63125	72011	78300
321 432	**SL**	H	*SL*	BY	78152	63126	72012	78301
321 433	**SL**	H	*SL*	BY	78153	63127	72013	78302
321 434	**SL**	H	*SL*	BY	78154	63128	72014	78303
321 435	**SL**	H	*SL*	BY	78155	63129	72015	78304
321 436	**SL**	H	*SL*	BY	78156	63130	72016	78305
321 437	**SL**	H	*SL*	BY	78157	63131	72017	78306
321 438	**GE**	H	*GE*	IL	78158	63132	72018	78307
321 439	**GE**	H	*GE*	IL	78159	63133	72019	78308
321 440	**GE**	H	*GE*	IL	78160	63134	72020	78309
321 441	**GE**	H	*GE*	IL	78161	63135	72021	78310
321 442	**GE**	H	*GE*	IL	78162	63136	72022	78311
321 443	**GE**	H	*GE*	IL	78125	63099	71985	78274
321 444	**GE**	H	*GE*	IL	78126	63100	71986	78275
321 445	**GE**	H	*GE*	IL	78127	63101	71987	78276
321 446	**GE**	H	*GE*	IL	78128	63102	71988	78277
321 447	**GE**	H	*GE*	IL	78129	63103	71989	78278
321 448	**GE**	H	*GE*	IL	78130	63104	71990	78279

Names (carried on ATS):

321 407 Hertfordshire WRVS
321 439 Chelmsford Cathedral Festival
321 444 Essex Lifeboats

Class 321/9. DTS(A)–PMS–ATS–DTS(B). Leased by West Yorkshire PTE from International Bank of Scotland. Managed by Porterbrook Leasing Company.

63153–63155. PMS. Dia. EC216. Lot No. 31109 BREL York 1991. –/79. 51.5 t.
72128–72130. ATS. Dia. EH240. Lot No. 31110 BREL York 1991. –/74 2T. 28.0 t.
77990–77992. DTS(A). Dia. EE277. Lot No. 31108 BREL York 1991. –/78. 29.3 t.
77993–77995. DTS(B). Dia. EE277. Lot No. 31111 BREL York 1991. –/78. 29.1 t.

321 901	**WY**	P	*NS*	NL	77990	63153	72128	77993
321 902	**WY**	P	*NS*	NL	77991	63154	72129	77994
321 903	**WY**	P	*NS*	NL	77992	63155	72130	77995

CLASS 322 4-Car Unit

DTC–PMS–ATS–DTS. Gangwayed within unit. Disc brakes.
Traction Motors: Four Brush TM2141C of 268 kW each.
Dimensions: 19.95 (DTC & DTS) or 19.92 (ATS & PMS) x 2.82 x 3.78 m.
Maximum Speed: 100 mph. **Doors:** Power operated sliding.
Couplers: Tightlock. **Bogies:** BREL P7-4/T3-7.
Multiple Working: Classes 313–323.
Non-Standard Livery:
- 322 481–483 carry 'Stansted Skytrain' livery (grey with a yellow stripe).

Advertising Livery:
- 322 484/485 'Stansted Express'.

78163–78167. DTC. Dia. EE313. Lot No. 31094 BREL York 1990. 35/22. 30.4 t.
63137–63141. PMS. Dia. EC215. Lot No. 31092 BREL York 1990. –/70. 52.3 t.
72023–72027. ATS. Dia. EH239. Lot No. 31093 BREL York 1990. –/60 2T. 29.5 t.
77985–77989. DTS. Dia. EE242. Lot No. 31091 BREL York 1990. –/65. 29.8 t.

322 481	**0**	H	*WN*	HE	78163	72023	63137	77985
322 482	**0**	H	*WN*	HE	78164	72024	63138	77986
322 483	**0**	H	*WN*	HE	78165	72025	63139	77987
322 484	**AL**	H	*WN*	HE	78166	72026	63140	77988
322 485	**AL**	H	*WN*	HE	78167	72027	63141	77989

CLASS 323 3-Car Unit

DMS(A)–PTS–DMS(B). Gangwayed within unit. Disc brakes.
Traction Motors: Four Holec DMKT 52/24 of 146 kW per power car.
Dimensions: 23.37 (DMS) or 23.44 (PTS) x 2.80 x . m.
Maximum Speed: 90 mph. **Doors:** Power operated sliding plug.
Couplers: Tightlock. **Bogies:** RFS BP62/BT52.
Multiple Working: Classes 313–323. **Notes:** 65003" is actually 65005.

64001–64043. DMS(A). Dia. EA272. Lot No. 31112 Hunslet TPL 1992–93. –/98 (* –/82). 41.0 t.
72201–72243. PTS. Dia. EH296. Lot No. 31113 Hunslet TPL 1992–93. –/88 (* –/80) 1T. 39.4 t.
65001–65043. DMS(B). Dia. EA272. Lot No. 31114 Hunslet TPL 1992–93. –/98 (* –/82). 41.0 t.

323 201	**CO**	P	*CT*	SI	64001	72201	65001
323 202	**CO**	P	*CT*	SI	64002	72202	65002
323 203	**CO**	P	*CT*	SI	64003	72203	65003"
323 204	**CO**	P	*CT*	SI	64004	72204	65004
323 205	**CO**	P	*CT*	SI	64005	72205	65003
323 206	**CO**	P	*CT*	SI	64006	72206	65006
323 207	**CO**	P	*CT*	SI	64007	72207	65007
323 208	**CO**	P	*CT*	SI	64008	72208	65008
323 209	**CO**	P	*CT*	SI	64009	72209	65009
323 210	**CO**	P	*CT*	SI	64010	72210	65010
323 211	**CO**	P	*CT*	SI	64011	72211	65011

323 212	**CO**	P	*CT*	SI	64012	72212	65012
323 213	**CO**	P	*CT*	SI	64013	72213	65013
323 214	**CO**	P	*CT*	SI	64014	72214	65014
323 215	**CO**	P	*CT*	SI	64015	72215	65015
323 216	**CO**	P	*CT*	SI	64016	72216	65016
323 217	**CO**	P	*CT*	SI	64017	72217	65017
323 218	**CO**	P	*CT*	SI	64018	72218	65018
323 219	**CO**	P	*CT*	SI	64019	72219	65021
323 220	**CO**	P	*CT*	SI	64020	72220	65020
323 221	**CO**	P	*CT*	SI	64021	72221	65019
323 222	**CO**	P	*CT*	SI	64022	72222	65022
323 223 *	**GM**	P	*NW*	LG	64023	72223	65023
323 224 *	**NW**	P	*NW*	LG	64024	72224	65024
323 225 *	**GM**	P	*NW*	LG	64025	72225	65025
323 226	**GM**	P	*NW*	LG	64026	72226	65026
323 227	**GM**	P	*NW*	LG	64027	72227	65027
323 228	**GM**	P	*NW*	LG	64028	72228	65028
323 229	**GM**	P	*NW*	LG	64029	72229	65029
323 230	**GM**	P	*NW*	LG	64030	72230	65030
323 231	**GM**	P	*NW*	LG	64031	72231	65031
323 232	**GM**	P	*NW*	LG	64032	72232	65032
323 233	**NW**	P	*NW*	LG	64033	72233	65033
323 234	**GM**	P	*NW*	LG	64034	72234	65034
323 235	**GM**	P	*NW*	LG	64035	72235	65035
323 236	**GM**	P	*NW*	LG	64036	72236	65036
323 237	**GM**	P	*NW*	LG	64037	72237	65037
323 238	**GM**	P	*NW*	LG	64038	72238	65038
323 239	**GM**	P	*NW*	LG	64039	72239	65039
323 240	**CO**	P	*CT*	SI	64040	72340	65040
323 241	**CO**	P	*CT*	SI	64041	72341	65041
323 242	**CO**	P	*CT*	SI	64042	72342	65042
323 243	**CO**	P	*CT*	SI	64043	72343	65043

CLASS 325 — 4-Car Royal Mail Unit

DTV(A)–PMV–TAV–DTV(B). Non gangwayed. Disc brakes.
Supply System: 25 kV 50 Hz a.c. overhead or 750 V d.c. third rail.
Traction Motors: Four GEC G315BZ of 247.5 kW each.
Dimensions: 20.35 x 2.82 x . m.
Doors: Roller shutter.
Maximum Speed: 100 mph.
Bogies: ABB P7-4/T3-7.
Couplers: Buckeye.
Multiple Working: Within class only.

68300–68330 (Even numbers). DTV(A). Dia. EE503. Lot No. 31144 ABB Derby 1995. Load capacity 12.0 t. 29.2 t.
68340–68355. PMV. Dia. EC501. Lot No. 31145 ABB Derby 1995. Load capacity 12.0 t. 49.5 t.
68360–68375. TAV. Dia. EH501. Lot No. 31146 ABB Derby 1995. Load capacity 12.0 t. 30.7 t.
68301–68331 (Odd numbers). DTV(B). Dia. EE503. Lot No. 31144 ABB Derby 1995. Load capacity 12.0 t. 29.1 t.

325 001	**RM**	RM	*E*	CE	68300	68340	68360	68301
325 002	**RM**	RM	*E*	CE	68302	68341	68361	68303
325 003	**RM**	RM	*E*	CE	68304	68342	68362	68305
325 004	**RM**	RM	*E*	CE	68306	68343	68363	68307
325 005	**RM**	RM	*E*	CE	68308	68344	68364	68309
325 006	**RM**	RM	*E*	CE	68310	68345	68365	68311
325 007	**RM**	RM	*E*	CE	68312	68346	68366	68313
325 008	**RM**	RM	*E*	CE	68314	68347	68367	68315
325 009	**RM**	RM	*E*	CE	68316	68348	68368	68317
325 010	**RM**	RM	*E*	CE	68318	68349	68369	68319
325 011	**RM**	RM	*E*	CE	68320	68350	68370	68321
325 012	**RM**	RM	*E*	CE	68322	68351	68371	68323
325 013	**RM**	RM	*E*	CE	68324	68352	68372	68325
325 014	**RM**	RM	*E*	CE	68326	68353	68373	68327
325 015	**RM**	RM	*E*	CE	68328	68354	68374	68329
325 016	**RM**	RM	*E*	CE	68330	68355	68375	68331

Names (carried on one DTV per side):

325 002 Royal Mail North Wales and North West
325 006 John Grierson
325 008 Peter Howarth C.B.E.

CLASS 332 Siemens 4-Car Express Unit

Various formations, see below. Gangwayed within unit. Disc brakes. Air conditioned.
Traction Motors: Two Siemens monomotors of 350 kW each per power car.
Dimensions: 23.74 x 2.75 x . m.
Maximum Speed: 160 km/h.
Doors: Power operated sliding plug.
Couplers: Scharfenberg.
Bogies: CAF design.
Multiple Working: Within Class only.

Units 332 001–332 007. DMF–TS–PTS–DMS.

63400–63406. PTS. Dia. EH243. CAF 1997–98. –/44 1T 1W. 45.6 t.
72400–72413. TS. Dia. EH245. CAF 1997–98. –/56. 45.2 t.
78400–78412 (Even numbers). DMF. CAF 1997–98. Converted 1998 from DMS. 26/–. 48.8 t.
78401–78413 (Odd numbers). DMS. Dia. EA243. CAF 1997–98. –/48. 48.8 t.

332 001	**HX**	HX	*HX*	OH	78400	72412	63400	78401
332 002	**HX**	HX	*HX*	OH	78402	72409	63401	78403
332 003	**HX**	HX	*HX*	OH	78404	72407	63402	78405
332 004	**HX**	HX	*HX*	OH	78406	72405	63403	78407
332 005	**HX**	HX	*HX*	OH	78408	72411	63404	78409
332 006	**HX**	HX	*HX*	OH	78410	72410	63405	78411
332 007	**HX**	HX	*HX*	OH	78412	72401	63406	78413

Units 332 008–332 014. DMS–TS–PTS–DMF.

63407–63413. PTS. Dia. EH243. CAF 1997–98. –/44 1T 1W. 45.6 t.
72400–72413. TS. Dia. EH245. CAF 1997–98. –/56. 45.2 t.
78414–78426 (Even numbers). DMS. Dia. EA244. CAF 1997–98. –/48. 48.8 t.
78415–78427 (Odd numbers). DMF. Dia. EA1??. CAF 1997–98. Converted 1998 from DMS. 14/– 1W. 48.8 t.

332 008	**HX**	HX	*HX*	OH	78414	72413	63407	78415
332 009	**HX**	HX	*HX*	OH	78416	72400	63408	78417
332 010	**HX**	HX	*HX*	OH	78418	72402	63409	78419
332 011	**HX**	HX	*HX*	OH	78420	72403	63410	78421
332 012	**HX**	HX	*HX*	OH	78422	72404	63411	78423
332 013	**HX**	HX	*HX*	OH	78424	72408	63412	78425
332 014	**HX**	HX	*HX*	OH	78426	72406	63413	78427

CLASS 333 Siemens 3-Car Unit

DMS(A)–TS–DMS(B). Gangwayed within unit. Disc brakes. Air conditioned.
Traction Motors: Two Siemens monomotors of ??? kW each per power car.
Dimensions: 23.74 x 2.75 x . m.
Maximum Speed: 160 km/h.
Doors: Power operated sliding plug.
Couplers: Scharfenberg.
Bogies: CAF design.
Multiple Working: Within Class only.

74461–74476. PTS. Dia. EH2??. CAF 2000.
78451–78481 (Odd numbers). DMS(A). Dia. EA2??. CAF 2000.
78452–78482 (Even numbers). DMS(B). Dia. EA2??. CAF 2000.

333 001	A	*NS*	78451	74461	78452
333 002	A	*NS*	78453	74462	78454
333 003	A	*NS*	78455	74463	78456
333 004	A	*NS*	78457	74464	78458
333 005	A	*NS*	78459	74465	78460
333 006	A	*NS*	78461	74466	78462
333 007	A	*NS*	78463	74467	78464
333 008	A	*NS*	78465	74468	78466
333 009	A	*NS*	78467	74469	78468
333 010	A	*NS*	78469	74470	78470
333 011	A	*NS*	78471	74471	78472
333 012	A	*NS*	78473	74472	78474
333 013	A	*NS*	78475	74473	78476
333 014	A	*NS*	78477	74474	78478
333 015	A	*NS*	78479	74475	78480
333 016	A	*NS*	78481	74476	78482

CLASS 334 Adtranz 'Juniper' 3-Car Unit

DMS(A)–PTS–DMS(B). Gangwayed within unit. Disc brakes. Air conditioned.
Traction Motors: Two Alstom Onix 800 of 270 kW each per power car.
Dimensions: 21.16 (DMS) or 19.94 (PTS) x 2.80 x . m.
Doors: Power operated sliding plug.
Maximum Speed: 100 mph. **Bogies:** Alstom LTB3/TBP3.
Couplers: Scharfenberg. **Multiple Working:** Within Class only.

64101–64140. DMS(A). Dia. EA215. Alstom Birmingham 1999–2000. –/64 1TD 2W. 42.6 t.
65101–65140. DMS(B). Dia. EA215. Alstom Birmingham 1999–2000. –/64. 42.6 t.
74301–74340. PTS. Dia.EH255. Alstom Birmingham 1999–2000. –/55. 39.4 t.

334 001	**CC**	H	*SR*	GW	64101	74301	65101
334 002	**CC**	H	*SR*	GW	64102	74302	65102
334 003	**CC**	H	*SR*	GW	64103	74303	65103
334 004		H	*SR*		64104	74304	65104
334 005		H	*SR*		64105	74305	65105
334 006		H	*SR*		64106	74306	65106
334 007		H	*SR*		64107	74307	65107
334 008		H	*SR*		64108	74308	65108
334 009		H	*SR*		64109	74309	65109
334 010		H	*SR*		64110	74310	65110
334 011		H	*SR*		64111	74311	65111
334 012		H	*SR*		64112	74312	65112
334 013		H	*SR*		64113	74313	65113
334 014		H	*SR*		64114	74314	65114
334 015		H	*SR*		64115	74315	65115
334 016		H	*SR*		64116	74316	65116
334 017		H	*SR*		64117	74317	65117
334 018		H	*SR*		64118	74318	65118
334 019		H	*SR*		64119	74319	65119
334 020		H	*SR*		64120	74320	65120
334 021		H	*SR*		64121	74321	65121
334 022		H	*SR*		64122	74322	65122
334 023		H	*SR*		64123	74323	65123
334 024		H	*SR*		64124	74324	65124
334 025		H	*SR*		64125	74325	65125
334 026		H	*SR*		64126	74326	65126
334 027		H	*SR*		64127	74327	65127
334 028		H	*SR*		64128	74328	65128
334 029		H	*SR*		64129	74329	65129
334 030		H	*SR*		64130	74330	65130
334 031		H	*SR*		64131	74331	65131
334 032		H	*SR*		64132	74332	65132
334 033		H	*SR*		64133	74333	65133
334 034		H	*SR*		64134	74334	65134
334 035		H	*SR*		64135	74335	65135
334 036		H	*SR*		64136	74336	65136
334 037		H	*SR*		64137	74337	65137

334 038	H	*SR*	64138	74338	65138
334 039	H	*SR*	64139	74339	65139
334 040	H	*SR*	64140	74340	65140

CLASS 357 Adtranz 'Electrostar' 4-Car Unit

DMS(A)–PTS–MS–DMS(B). Gangwayed within unit. Disc and regenerative brakes. Air conditioning.
Supply System: 25 kV a.c. 50Hz overhead (with provision for 750 V d.c. third rail supply). **Bogies**: Adtranz P3-25/ T3-25.
Traction Motors: Two Adtranz of 250 kW each per motor car.
Dimensions: 20.40 (DMS) or 19.99 (MS & PTS) x 2.80 x 3.78 m.
Maximum Speed: 100 mph. **Doors**: Power operated sliding plug.
Multiple Working: Within class. **Couplers**: Tightlock.

67651–67694. DMS(A). Dia. EA273. Adtranz Derby 1999–2000. –/71. 40.7 t.
67751–67794. DMS(B). Dia. EA214. Adtranz Derby 1999–2000. –/71. 40.7 t.
74051–74094. PTS. Dia. EH215. Adtranz Derby 1999–2000. –/62 1TD 2W. 36.7 t.
74151–74194. MS. Dia. EC225. Adtranz Derby 1999–2000. –/78 . 39.5 t.

357 001	**LS**	P	*LS*	EM	67651	74051	74151	67751
357 002	**LS**	P	*LS*	EM	67652	74052	74152	67752
357 003		P	*LS*		67653	74053	74153	67753
357 004		P	*LS*		67654	74054	74154	67754
357 005		P	*LS*		67655	74055	74155	67755
357 006		P	*LS*		67656	74056	74156	67756
357 007		P	*LS*		67657	74057	74157	67757
357 008		P	*LS*		67658	74058	74158	67758
357 009		P	*LS*		67659	74059	74159	67759
357 010		P	*LS*		67660	74060	74160	67760
357 011		P	*LS*		67661	74061	74161	67761
357 012		P	*LS*		67662	74062	74162	67762
357 013		P	*LS*		67663	74063	74163	67763
357 014		P	*LS*		67664	74064	74164	67764
357 015		P	*LS*		67665	74065	74165	67765
357 016		P	*LS*		67666	74066	74166	67766
357 017		P	*LS*		67667	74067	74167	67767
357 018		P	*LS*		67668	74068	74168	67768
357 019		P	*LS*		67669	74069	74169	67769
357 020		P	*LS*		67670	74070	74170	67770
357 021		P	*LS*		67671	74071	74171	67771
357 022		P	*LS*		67672	74072	74172	67772
357 023		P	*LS*		67673	74073	74173	67773
357 024		P	*LS*		67674	74074	74174	67774
357 025		P	*LS*		67675	74075	74175	67775
357 026		P	*LS*		67676	74076	74176	67776
357 027		P	*LS*		67677	74077	74177	67777
357 028		P	*LS*		67678	74078	74178	67778
357 029		P	*LS*		67679	74079	74179	67779
357 030		P	*LS*		67680	74080	74180	67780
357 031		P	*LS*		67681	74081	74181	67781

357 032	P	*LS*	67682	74082	74182	67782
357 033	P	*LS*	67683	74083	74183	67783
357 034	P	*LS*	67684	74084	74184	67784
357 035	P	*LS*	67685	74085	74185	67785
357 036	P	*LS*	67686	74086	74186	67786
357 037	P	*LS*	67687	74087	74187	67787
357 038	P	*LS*	67688	74088	74188	67788
357 039	P	*LS*	67689	74089	74189	67789
357 040	P	*LS*	67690	74090	74190	67790
357 041	P	*LS*	67691	74091	74191	67791
357 042	P	*LS*	67692	74092	74192	67792
357 043	P	*LS*	67693	74093	74193	67793
357 044	P	*LS*	67694	74094	74194	67794

CLASS 365 'Networker' 4-Car Express Unit

DMC(A)–TSD–TS–DMC(B). Gangwayed within unit. Disc, rheostatic and regenerative braking.
Supply System: 25 kV a.c. 50 Hz overhead with provision for 750 V d.c. third rail supply or 750 V d.c. third rail with provision for 25 kV a.c. 50 Hz overhead supply.
Traction Motors: Four GEC-Alsthom G354CX of 157 kW each per power car.
Dimensions: 20.89 (DMC) or 20.06 (TS & TSD) x 2.81 x 3.77 m.
Maximum Speed: 100 mph. **Doors:** Power operated sliding plug.
Couplers: Tightlock. **Bogies:** ABB P7-4/T3-16.
Multiple Working: Classes 365, 465 and 466.

65894–65934. DMC(A). Dia. EA301. Lot No. 31133 ABB York 1994–95. 12/56. 41.7 t.
65935–65975. DMC(B). Dia. EA301. Lot No. 31136 ABB York 1994–95. 12/56. 41.7 t.
72240–72320 (Even nos.). PTS. Dia. EH298. Lot No. 31135 ABB York 1994–95. –/68 1T. 34.6 t.
72241–72321 (Odds). TSD. Dia. EH298. Lot No. 31134 ABB York 1994–95. –/65 1TD 1W. 32.9 t.

Units 365 001–365 516. 750 V d.c. third rail with provision for 25 kV a.c. 50 Hz overhead supply.

365 501	**CS**	H	*SE*	SG	65894	72241	72240	65935
365 502	**CS**	H	*SE*	SG	65895	72243	72242	65936
365 503	**CS**	H	*SE*	SG	65896	72245	72244	65937
365 504	**CS**	H	*SE*	SG	65897	72247	72246	65938
365 505	**CS**	H	*SE*	SG	65898	72249	72248	65939
365 506	**CS**	H	*SE*	SG	65899	72251	72250	65940
365 507	**CS**	H	*SE*	SG	65900	72253	72252	65941
365 508	**CS**	H	*SE*	SG	65901	72255	72254	65942
365 509	**CS**	H	*SE*	SG	65902	72257	72256	65943
365 510	**CS**	H	*SE*	SG	65903	72259	72258	65944
365 511	**CS**	H	*SE*	SG	65904	72261	72260	65945
365 512	**CS**	H	*SE*	SG	65905	72263	72262	65946
365 513	**CS**	H	*SE*	SG	65906	72265	72264	65947
365 514	**CS**	H	*SE*	SG	65907	72267	72266	65948

365 515	**CS**	H	*SE*	SG	65908	72269	72268	65949
365 516	**CS**	H	*SE*	SG	65909	72271	72270	65950

Units 365 517–365 541. 25 kV a.c. 50 Hz overhead with provision for 750 V d.c. third rail supply.

365 517	**NT**	H	*WN*	HE	65910	72273	72272	65951
365 518	**NT**	H	*WN*	HE	65911	72275	72274	65952
365 519	**NT**	H	*WN*	HE	65912	72277	72276	65953
365 520	**NT**	H	*WN*	HE	65913	72279	72278	65954
365 521	**NT**	H	*WN*	HE	65914	72281	72280	65955
365 522	**NT**	H	*WN*	HE	65915	72283	72282	65956
365 523	**NT**	H	*WN*	HE	65916	72285	72284	65957
365 524	**NT**	H	*WN*	HE	65917	72287	72286	65958
365 525	**NT**	H	*WN*	HE	65918	72289	72288	65959
365 526	**NT**	H	*WN*	HE	65919	72291	72290	65960
365 527	**NT**	H	*WN*	HE	65920	72293	72292	65961
365 528	**NT**	H	*WN*	HE	65921	72295	72294	65962
365 529	**NT**	H	*WN*	HE	65922	72297	72296	65963
365 530	**NT**	H	*WN*	HE	65923	72299	72298	65964
365 531	**NT**	H	*WN*	HE	65924	72301	72300	65965
365 532	**NT**	H	*WN*	HE	65925	72303	72302	65966
365 533	**NT**	H	*WN*	HE	65926	72305	72304	65967
365 534	**NT**	H	*WN*	HE	65927	72307	72306	65968
365 535	**NT**	H	*WN*	HE	65928	72309	72308	65969
365 536	**NT**	H	*WN*	HE	65929	72311	72310	65970
365 537	**NT**	H	*WN*	HE	65930	72313	72312	65971
365 538	**NT**	H	*WN*	HE	65931	72315	72314	65972
365 539	**NT**	H	*WN*	HE	65932	72317	72316	65973
365 540	**NT**	H	*WN*	HE	65933	72319	72318	65974
365 541	**NT**	H	*WN*	HE	65934	72321	72320	65975

Names (carried on one side of each DMC):

365 505 Spirit of Ramsgate | 365 515 Spirit of Dover

CLASS 375 Adtranz 'Electrostar' 3- or 4-Car Unit

Various formations, see below. Gangwayed throughout. Disc and regenerative braking. Air conditioned.
Supply System: 25 kV a.c. 50 Hz overhead and/or 750 V d.c. third rail. 750 V d.c only units also have provision for 25 kV a.c. 50 Hz overhead supply.
Traction Motors: Two Adtranz of 250 kW each per motor car.
Dimensions: 20.40 (DMS) or 19.99 (MS & PTS) x 2.80 x 3.78 m.
Maximum Speed: 100 mph.
Doors: Power operated sliding plug.
Couplers: Tightlock.
Multiple Working: Within class.
Bogies: Adtranz P3-25/T3-25.

Class 375/3. 750 V d.c. only 3-car units. DMS(A)–PTS–DMS(B).

67921–67930. DMS(A). Dia. EA277. Adtranz Derby 2000. –/64. . t.
67931–67940. DMS(B). Dia. EA277. Adtranz Derby 2000. –/64. . t.
74351–74360. PTS. Dia. EH254. Adtranz Derby 2000. –/56 1TD 2W. . t.

375 301		H	*SE*		67921	74351	67931	
375 302		H	*SE*		67922	74352	67932	
375 303		H	*SE*		67923	74353	67933	
375 304		H	*SE*		67924	74354	67934	
375 305		H	*SE*		67925	74355	67935	
375 306		H	*SE*		67926	74356	67936	
375 307		H	*SE*		67927	74357	67937	
375 308		H	*SE*		67928	74358	67938	
375 309		H	*SE*		67929	74359	67939	
375 310		H	*SE*		67930	74360	67940	

Class 375/6. Dual Voltage 4-car units. DMS(A)–PTS–MS–DMS(B).

67801–67830. DMS(A). Dia. EA275. Adtranz Derby 1999–2000. –/60. 46.2 t.
67851–67880. DMS(B). Dia. EA275. Adtranz Derby 1999–2000. –/60. 46.2 t.
74201–74230. PTS. Dia. EH252. Adtranz Derby 1999–2000. –/56 1TD 2W. 40.7 t.
74251–74280. MS. Dia. EC230. Adtranz Derby 1999–2000. –/66 1T. 40.5 t.

375 601	**U**	H	*SE*	AF	67801	74201	74251	67851
375 602		H	*SE*		67802	74202	74252	67852
375 603		H	*SE*		67803	74203	74253	67853
375 604		H	*SE*		67804	74204	74254	67854
375 605		H	*SE*		67805	74205	74255	67855
375 606		H	*SE*		67806	74206	74256	67856
375 607		H	*SE*		67807	74207	74257	67857
375 608		H	*SE*		67808	74208	74258	67858
375 609		H	*SE*		67809	74209	74259	67859
375 610		H	*SE*		67810	74210	74260	67860
375 611		H	*SE*		67811	74211	74261	67861
375 612		H	*SE*		67812	74212	74262	67862
375 613		H	*SE*		67813	74213	74263	67863
375 614		H	*SE*		67814	74214	74264	67864
375 615		H	*SE*		67815	74215	74265	67865
375 616		H	*SE*		67816	74216	74266	67866
375 617		H	*SE*		67817	74217	74267	67867
375 618		H	*SE*		67818	74218	74268	67868
375 619		H	*SE*		67819	74219	74269	67869
375 620		H	*SE*		67820	74220	74270	67870
375 621		H	*SE*		67821	74221	74271	67871
375 622		H	*SE*		67822	74222	74272	67872
375 623		H	*SE*		67823	74223	74273	67873
375 624		H	*SE*		67824	74224	74274	67874
375 625		H	*SE*		67825	74225	74275	67875
375 626		H	*SE*		67826	74226	74276	67876
375 627		H	*SE*		67827	74227	74277	67877
375 628		H	*SE*		67828	74228	74278	67878
375 629		H	*SE*		67829	74229	74279	67879
375 630		H	*SE*		67830	74230	74280	67880

Class 375/7. 750 V d.c. only 4-car units. DMS(A)–PTS–MS–DMS(B).

67831–67845. DMS(A). Dia. EA276. Adtranz Derby 1999–2000. –/60. . t.
67881–67895. DMS(B). Dia. EA276. Adtranz Derby 1999–2000. –/60. t.
74231–74245. PTS. Dia. EH253. Adtranz Derby 1999–2000. –/56 1TD 2W. . t.
74281–74295. MS. Dia. EC231. Adtranz Derby 1999–2000. –/66 1T. . t.

375 701	H	*SE*	67831	74281	74231	67881
375 702	H	*SE*	67832	74282	74232	67882
375 703	H	*SE*	67833	74283	74233	67883
375 704	H	*SE*	67834	74284	74234	67884
375 705	H	*SE*	67835	74285	74235	67885
375 706	H	*SE*	67836	74281	74236	67886
375 707	H	*SE*	67837	74287	74237	67887
375 708	H	*SE*	67838	74288	74238	67888
375 709	H	*SE*	67839	74289	74239	67889
375 710	H	*SE*	67840	74290	74240	67890
375 711	H	*SE*	67841	74291	74241	67891
375 712	H	*SE*	67842	74292	74242	67892
375 713	H	*SE*	67843	74293	74243	67893
375 714	H	*SE*	67844	74294	74244	67894
375 715	H	*SE*	67845	74295	74245	67895

2. 750 V d.c. THIRD RAIL UNITS

Supply System: 660–850 V d.c. third rail unless otherwise stated.

CLASS 483 2-Car Unit

DMS(A)–DMS(B). Non gangwayed. End doors. Converted from vehicles purchased from London Transport in 1988.
Traction Motors: Two Crompton Parkinson/GEC/BTH LT100 of 125 kW each per power car.
Dimensions: 15.94 x 2.65 x 2.88 m. **Doors**: Power operated sliding.
Maximum Speed: 45 mph. **Bogies**: LT design.
Couplings: Wedgelock. **Multiple Working**: Within class.

121–129. DMS(A). Dia. EA265. Met-Camm. 1938. Rebuilt to Lot No. 31071 BRML Eastleigh 1989–92. –/42. 27.5 t.
221–229. DMS(B). Dia. EA266. Met-Camm. 1938. Rebuilt to Lot No. 31072 BRML Eastleigh 1989–92. –/42. 27.5 t.

002	**N**	H	*IL*	RY	122	225
003	**N**	H		RY(S)	123	221
004	**N**	H	*IL*	RY	124	224
006	**N**	H	*IL*	RY	126	226
007	**N**	H	*IL*	RY	127	227
008	**N**	H	*IL*	RY	128	228
009	**N**	H	*IL*	RY	129	229

CLASS 438 3-TC Express Trailer Unit

DTS–TBS–DTS. Gangwayed throughout.
Dimensions: 20.18 x 2.82 x 3.81 m. **Doors**: Manually operated slam.
Maximum Speed: 90 mph. **Bogies**: B5 (SR).
Couplings: Buckeye. **Multiple Working**: SR type.

70812/826. TBS. Dia. EJ260. Lot No. 30765 York 1966–67. Rebuilt from loco-hauled BSK to Lot No. 30229 Met. Camm. 1955–57. –/32. 1T. 33.5 t.
76301–76302. DTS. Dia. EE266. Lot No. 30764 York 1966–67. Rebuilt from loco-hauled TSO to Lot No. 30219 Swindon 1955–57. –/64. 32.0 t.

417	**B**	CM	TM	76301	70826	76302
Spare	**B**	CM	TM		70812	

CLASSES 411 & 412 3- or 4-Cep/Bep Express Unit

Various formations, see below. Gangwayed throughout.
Traction Motors: Two English Electric 507 of 185 kW each per power car.
Dimensions: 20.34 x 2.82 x 3.83 m.
Maximum Speed: 90 mph. **Doors**: Manually operated slam.
Couplings: Buckeye. **Multiple Working**: SR type.
Bogies: Mk. 4 (* Mk. 3B; † Mk. 6)/Commonwealth († B5 (SR)).

61229–61239 (Odd numbers). DMS(A). Dia. EA264. Lot No. 30449 Eastleigh 1958. –/64. 44.2 t.
61230–61240 (Even numbers). DMS(B). Dia. EA264. Lot No. 30449 Eastleigh 1958. –/64. 43.5 t.
61305–61409 (Odd numbers). DMS(A). Dia. EA264. Lot No. 30454 Eastleigh 1958–59. –/64. 44.2 t.
61304–61408 (Even numbers). DMS(B). Dia. EA264. Lot No. 30454 Eastleigh 1958–59. –/64. 43.5 t.
61694–61810 (Even numbers). DMS(A). Dia. EA264. Lot No. 30619 Eastleigh 1960–61. –/64. 44.2 t.
61695–61811 (Odd numbers). DMS(B). Dia. EA264. Lot No. 30619 Eastleigh 1960–61. –/64. 43.5 t.
61868–61870 (Even numbers) DMS(A). Dia. EA264. Lot No. 30638 Eastleigh 1961. –/64. 44.2 t.
61869–61871 (Odd numbers) DMS(B). Dia. EA264. Lot No. 30638 Eastleigh 1961. –/64. 43.5 t.
61948–61960 (Even numbers). DMS(A). Dia. EA264. Lot No. 30708 Eastleigh 1963. –/64. 44.2 t.
61949–61961 (Odd numbers). DMS(B). Dia. EA264. Lot No. 30708 Eastleigh 1963. –/64. 43.5 t.
69341–69347. TRBS. Dia. EN261. Built as TRB to Lot No. 30622 Eastleigh 1961. Converted BREL Swindon 1982–84. –/24 plus 9 chairs 1T. 35.5 t.
70033–70036. TBC. Dia. EJ361. Lot No. 30109 Eastleigh 1956. 24/6 2T. 36.2 t.
70043–70044. TBC. Dia. EJ361. Lot No. 30639 Eastleigh 1961. 24/6 2T. 36.2 t.
70229–70234. TBC. Dia. EJ361. Lot No. 30450 Eastleigh 1958. 24/6 2T. 36.2 t.
70235–70240. TBC. Dia. EJ361. Lot No. 30451 Eastleigh 1958. 24/6 2T. 36.2 t.
70241–70242. TBC. Dia. EJ361. Lot No. 30640 Eastleigh 1961. 24/6 2T. 36.2 t.
70260–70302. TS. Dia. EH282. Lot No. 30455 Eastleigh 1958–59. –/64 2T. 31.5 t.
70303–70355. TS. Dia. EH282. Lot No. 30456 Eastleigh 1958–59. –/64 2T. 31.5 t.
70503–70551. TS. Dia. EH282. Lot No. 30620 Eastleigh 1960–61. –/64 2T. 31.5 t.
70552–70610. TS. Dia. EH282. Lot No. 30621 Eastleigh 1960–61. –/64 2T. 31.5 t.
70653–70659. TS. Dia. EH282. Lot No. 30709 Eastleigh 1963. –/64 2T. 31.5 t.
70660–70666. TS. Dia. EH282. Lot No. 30710 Eastleigh 1963. –/64 2T. 31.5 t.
71625–71636. TS. Dia. EH284. Converted BREL Swindon 1981–82 from loco-hauled TSO of various lots. –/64 2T. 33.6 t.
71711–71712. TS. Dia. EH284. Converted BREL Swindon 1983–84 from loco-hauled TSO of various lots. –/64 2T. 33.6 t.

Class 411/9. DMS(A)–TBC–DMS(B).

No.	*Former No.*								
1101	1530		**N**	P	*SE*	RM	61331	70316	61330
1102	1561		**N**	P	*SE*	RM	61231	70604	61232
1103	1610	*	**N**	P	*SE*	RM	61750	70580	61751
1104	1613	*	**N**	P	*SE*	RM	61760	70585	61761
1105	1619	*	**N**	P	*SE*	RM	61952	70655	61953
1106	1510		**N**	P	*SE*	RM	61365	70333	61364
1107	1520		**N**	P	*SE*	RM	61343	70327	61380
1108	1536		**N**	P	*SE*	RM	61399	70350	61398
1109	1541		**N**	P	*SE*	RM	61409	70355	61408
1110	1543		**N**	P	*SE*	RM	61323	70312	61322
1111	1549		**N**	P	*SE*	RM	61339	70320	61338

1112	1554		**N**	P	*SE*	RM	61369 70335	61368
1113	1556		**N**	P	*SE*	RM	61371 70336	61370
1114	1559		**N**	P	*SE*	RM	61377 70339	61376
1115	1577	*	**N**	P	*SE*	RM	61718 70564	61719
1116	1580	*	**N**	P	*SE*	RM	61756 70589	61757
1117	1595	*	**N**	P	*SE*	RM	61704 70557	61705
1118	1597	*	**N**	P	*SE*	RM	61708 70559	61709

(Class continued with 1507)

CLASSES 421 & 422 4-Cig or 3-Cop Express Unit

Various formations, see below. Gangwayed throughout.
Traction Motors: Four English Electric 507 of 185 kW each.
Dimensions: 20.19 x 2.82 x 3.86 m.
Maximum Speed: 90 mph.
Doors: Manually operated slam (d – central door locking).
Couplings: Buckeye. **Multiple Working**: SR type.
Bogies: Mk. 4 or Mk. 6/B5 (SR).

62017–62070. MBS. Dia. ED264. Lot No. 30742 York 1964–65. –/56. 49.0 t.
62277–62286. MBS. Dia. ED264. Lot No. 30804 York 1970. –/56. (§ 1W) 49.0 t.
62287–62316. MBS. Dia. ED 264. Lot No. 30808 York 1970. –/56. (§ 1W) 49.0 t.
62355–62425. MBS. Dia. ED264. Lot No. 30816 York 1970. –/56. (§ 1W) 49.0 t.
62430. MBS. Dia. ED264. Lot No. 30829 York 1972. –/56. 49.0 t.
69301–69318. TRBS. Dia. EN260. Lot No. 30744 York 1966. –/40. 35.0 t.
69333–69339. TRBS. Dia. EN260. Lot No. 30805 York 1970. –/40. 35.0 t.
70260–70302. TS. Dia. EH282. Lot No. 30455 Eastleigh 1958–59. –/64 2T. 31.5 t. Class 411/5 cars.
70503–70551. TS. Dia. EH282. Lot No. 30620 Eastleigh 1960–61. –/64 2T. 31.5 t. Class 411/5 cars.
70695–70730. TS. Dia. EH287. Lot No. 30730 York 1964–65. –/72. 31.5 t.
70967–70996 TS. Dia. EH287 (• EH275). Lot No. 30809 York 1970–71. –/72. 31.5t.
71035–71105. TS. Dia. EH287. Lot No. 30817 York 1970. –/72. 31.5t.
71106. TS. Dia. EH287. Lot No. 30830 York 1972. –/72. 31.5t.
71766–71770. TS. Dia. EH287. Built as EMU TSRB to Lot No. 30744 York 1963–66. Converted BRML Eastleigh 1985–87. –/72. 31.5 t.
71926. TS. Dia. EH287. Built as EMU TSRB to Lot No. 30744 York 1963–66. Converted BRML Eastleigh 1988. –/72. 31.5t.
71927–71928. TS. Dia. EH287. Built as EMU TSRB to Lot No. 30805 York 1970. Converted BRML Eastleigh 1988. –/72. 31.5t.
76022–76075. DTC(B). Dia. EE369. Lot No. 30740 York 1964–65. 18/36 2T.
76076–76129. DTC(A) (†‡ DTS). Dia. EE369 (†‡ EE282). Lot No. 30741 York 1964–65. 18/36 († –/54; ‡ –/60) 2T. 35.5 t.
76561–76570. DTC(A) (†‡ DTS; § DTS(A)). Dia. EE369 (†‡ EE283; § EE245). Lot No. 30802 York 1970 (§ Rebuilt Wessex Traincare/Alstom Eastleigh 1997–98). 18/36 (*12/42; †–/54; ‡§ –/60) 2(§ 1)T. 35.5 t.
76571–76580. DTC(B) (§ DTS(B)). Dia. EE369 (§ EE246). Lot No. 30802 York 1970 (§ Rebuilt Wessex Traincare/Alstom Eastleigh 1997–98). 18/36 2(§ 1)T. 35.5 t.

76581–76610. DTC(A) (†‡ DTS; § DTS(A)). Dia. EE369 (§ EE245). Lot No. 30806 York 1970 (§ Rebuilt Wessex Traincare/Alstom Eastleigh 1997–98). 18/36 (*12/42; † –/54; ‡§–/60) 2(§ 1)T. 35.5 t.
76611–76640. DTC(B) (§ DTS(B)). Dia. EE369 (§ EE246). Lot No. 30807 York 1970 (§ Rebuilt Wessex Traincare/Alstom Eastleigh 1997–98). 18/36 2(§1)T. 35.5 t.
76717–76787. DTC(A) (†‡ DTS; § DTS(A)). Dia. EE369 (†‡ EE282; § EE245). Lot No. 30814 York 1970–72 (§ Rebuilt Wessex Traincare/Alstom Eastleigh 1997–98). 18/36 (* 12/42; † –/54; ‡§ –/60) 2(§ 1)T. 35.5 t.
76788–76858. DTC(B) (§ DTS(B)). Dia. EE369 (§ EE246). Lot No. 30815 York 1970–72. (§ Rebuilt Wessex Traincare/Alstom Eastleigh 1997–98). 18/36 2T. 35.5 t.
76859. DTC(A). Dia. EE369. Lot No. 30827 York 1972. 12/42 2T. 35.5 t.
76860. DTC(B). Dia. EE369. Lot No. 30828 York 1972. 12/42 2T. 35.5 t.

Class 421/5. DTC(A)–MBS–TS–DTC(B). 'Greyhound' units with additional stage of field weakening to improve the maximum attainable speed. Mk. 6 motor bogies.

1301	**ST**	H	*SW*	FR	76595	62301	70981	76625
1302	**ST**	H	*SW*	FR	76584	62290	70970	76614
1303	**ST**	H	*SW*	FR	76581	62287	70967	76611
1304	**ST**	H	*SW*	FR	76583	62289	70969	76613
1305	**ST**	H	*SW*	FR	76717	62355	71035	76788
1306	**ST**	H	*SW*	FR	76723	62361	71041	76794
1307	**ST**	H	*SW*	FR	76586	62292	70972	76616
1308	**ST**	H	*SW*	FR	76627	62298	70978	76622
1309	**ST**	H	*SW*	FR	76594	62300	70980	76624
1310	**ST**	H	*SW*	FR	76567	62283	71926	76577
1311	**ST**	H	*SW*	FR	76561	62277	71927	76571
1312	**ST**	H	*SW*	FR	76562	62278	71928	76572
1313	**ST**	H	*SW*	FR	76596	62302	70982	76626
1314	**ST**	H	*SW*	FR	76588	62294	70974	76618
1315	**ST**	H	*SW*	FR	76608	62314	70994	76638
1316	**ST**	H	*SW*	FR	76585	62291	70971	76615
1317	**ST**	H	*SW*	FR	76597	62303	70983	76592
1318	**ST**	H	*SW*	FR	76590	62296	70976	76620
1319	**ST**	H	*SW*	FR	76591	62297	70977	76621
1320	**ST**	H	*SW*	FR	76593	62299	70979	76623
1321	**ST**	H	*SW*	FR	76589	62295	70975	76619
1322	**ST**	H	*SW*	FR	76587	62293	70973	76617

Class 421/8. DTC(A)–MBS–TS–DTC(B) 'Greyhound' units with additional stage of field weakening to improve the maximum attainable speed. Mk. 6 motor bogies. Former Class 422 units with TRBS replaced by Class 411/5 TS.

No.	*Former No.*								
1392	2255	**ST**	P	*SW*	FR	76811	62378	70273	76740
1393	2258	**ST**	P	*SW*	FR	76746	62384	70527	76817
1394	2251	**ST**	P	*SW*	FR	76726	62364	70663	76797
1395	2262	**ST**	P	*SW*	FR	76850	62417	70662	76779
1396	2254	**ST**	P	*SW*	FR	76803	62370	70531	76732
1397	2260	**ST**	P	*SW*	FR	76749	62387	70515	76820
1398	2259	**ST**	P	*SW*	FR	76819	62386	70292	76748
1399	2256	**ST**	P	*SW*	FR	76747	62385	70508	76818

Class 421/7. DTS(A)–MBS–DTS(B). 3-car units for Connex South Central Brighton–Portsmouth 'Coastway' route. Mk. 6 motor bogies.

1401	§	**CX**	P	*SC*	BI	76568	62284	76578
1402	§	**CX**	P	*SC*	BI	76564	62280	76574
1403	§	**CX**	P	*SC*	BI	76563	62279	76573
1404	§	**CX**	P	*SC*	BI	76602	62308	76632
1405	§	**CX**	P	*SC*	BI	76565	62281	76575
1406	§	**CX**	P	*SC*	BI	76728	62366	76799
1407	§	**CX**	P	*SC*	BI	76729	62367	76800
1408	§	**CX**	P	*SC*	BI	76750	62388	76821
1409	§	**CX**	P	*SC*	BI	76569	62285	76579
1410	§	**CX**	P	*SC*	BI	76734	62372	76805
1411	§	**CX**	P	*SC*	BI	76570	62286	76580

Name (Carried on MBS):

1409 Operation Perseus

(Class continued with 1701)

CLASS 411 4-Cep Express Unit

For details see pages 54–55.

Class 411/5. 4-car units. DMS(A)–TBC–TS–DMS(B).

1507	**ST**	P	*SW*	FR	61363	70332	70289	61362
1509	**N**	P	*SE*	RM	61335	70318	70275	61334
1511	**N**	P	*SE*	RM	61367	70334	70291	61366
1512	**ST**	P	*SW*	FR	61321	70311	70268	61320
1517	**N**	P	*SW*	FR	61317	70309	70266	61316
1518	**N**	P		ZG(S)	61333	70317	70274	61332
1519	**ST**	P	*SW*	FR	61403	70352	70516	61402
1527	**N**	P	*SW*	FR	61237	70239	70233	61238
1531	**ST**	P	*SW*	FR	61233	70237	70231	61234
1532	**N**	P		FR(S)	61391	70346	71628	61390
1533	**ST**	P	*SW*	FR	61393	70347	71627	61385
1534	**ST**	P	*SW*	FR	61405	70353	71626	61404
1535	**ST**	P	*SW*	FR	61397	70349	71629	61396
1537	**ST**	P	*SW*	FR	61229	70235	70229	61230
1538	**ST**	P	*SW*	FR	61307	70304	70261	61306
1539	**ST**	P	*SW*	FR	61401	70351	71632	61400
1544	**ST**	P	*SW*	FR	61315	70308	70265	61349
1547	**ST**	P	*SW*	FR	61329	70578	70272	61328
1548	**ST**	P	*SW*	FR	61375	70338	70295	61374
1550	**N**	P	*SW*	FR	61313	70307	70264	61312
1551	**N**	P	*SE*	RM	61325	70313	70270	61324
1553	**N**	P	*SW*	FR	61728	70306	70263	61350
1555	**N**	P	*SW*	FR	61311	70326	70283	61310
1557	**N**	P	*SW*	FR	61337	70331	70288	61360
1560	**N**	P	*SE*	RM	61387	70344	70301	61386
1562	**N**	P	*SE*	RM	61407	70236	70241	61406

1563	*	**N**	P	*SW*	FR	61740	70575	70526	61741
1564	*	**N**	P	*SE*	RM	61788	70599	70550	61789
1565	*	**ST**	P	*SW*	FR	61762	70586	71711	61763
1566	*	**ST**	P	*SW*	FR	61722	70566	70517	61723
1568	*	**ST**	P	*SW*	FR	61766	70588	70539	61767
1570	*	**N**	P	*SE*	RM	61738	70574	70525	61739
1571	*	**N**	P	*SW*	FR	61806	70608	71636	61807
1572	*	**N**	P	*SW*	FR	61734	70572	70523	61735
1573	*	**ST**	P	*SW*	FR	61726	70568	70519	61727
1574	*	**N**	P	*SE*	RM	61792	70601	71635	61793
1575	*	**N**	P	*SE*	RM	61768	70583	70540	61769
1576	*	**N**	P	*SE*	RM	61770	70590	70541	61771
1578	*	**ST**	P	*SW*	FR	61700	70555	70506	61701
1581	*	**ST**	P	*SW*	FR	61784	70597	70548	61785
1582	*	**N**	P	*SE*	RM	61748	70603	71630	61797
1584	*	**N**	P	*SE*	RM	61752	70581	70532	61753
1585	*	**N**	P	*SE*	RM	61710	70560	70511	61711
1586	*	**N**	P	*SE*	RM	61714	70562	70513	61715
1587	*	**N**	P	*SE*	RM	61764	70587	71625	61765
1588	*	**N**	P	*SE*	RM	61720	70044	70520	61721
1589	*	**N**	P		ZG(S)	61742	70576		61743
1590	*	**N**	P	*SE*	RM	61696	70553	70504	61697
1591	*	**N**	P	*SE*	RM	61790	70600	70551	61791
1592	*	**N**	P	*SE*	RM	61778	70594	70545	61779
1593	*	**N**	P	*SE*	RM	61730	70570	70521	61731
1594	*	**N**	P	*SE*	RM	61754	70582	70533	61755
1599	*	**N**	P	*SE*	RM	61706	70558	70509	61707
1602	*	**CX**	P	*SE*	RM	61958	70565	70279	61959
1606	*	**N**	P		AF(S)	61694	70552		61695
1607	*	**N**	P	*SE*	RM	61698	70554	70505	61699
1609	*	**N**	P	*SE*	RM	61744	70577	70528	61745
1611	*	**N**	P	*SE*	RM	61758	70584	70537	61759
1612	*	**ST**	P	*SW*	FR	61794	70602	70535	61795
1614	*	**N**	P	*SE*	RM	61702	70556	70507	61703
1615	*	**N**	P	*SE*	RM	61956	70657	70664	61957
1616	*	**N**	P	*SE*	RM	61950	70654	70543	61951
1617	*	**N**	P	*SW*	FR	61800	70605	70661	61801
1618	*	**N**	P		RM(S)	61868	70043	70230	61869
1697	†	**N**	P	*SW*	FR	61373	70337	70294	61372
1698	†	**N**	P	*SW*	FR	61355	70343	70300	61384
1699	†	**N**	P	*SW*	FR	61712	70561	70512	61713
Spare		**N**	P		ZD(S)	61383			

Spare TS.

Spare	**N**	P	SU(S)	70035			
Spare	**N**	P	ZG(S)	70503	70536	70663	
Spare	**N**	P	CJ(S)	70277	70284	70290	70293
Spare	**N**	P	CJ(S)	70296	70297	70510	71631
Spare	**N**	P	CJ(S)	71633			

CLASS 421 4-Cig Express Unit

For details see pages 56–57.

Class 421/3. DTC(A)/(†‡DTS)–MBS–TS–DTC(B). Mk. 4 motor bogies.

1701		**U**	A	*SE*	RM	76087	62028	70706	76033
1702	†	**CX**	A	*SC*	BI	76101	62042	70720	76047
1703	†	**CX**	A	*SC*	BI	76097	62038	70716	76043
1704	†	**CX**	A	*SC*	BI	76092	62033	70711	76038
1705	†	**CX**	A	*SC*	BI	76076	62017	70695	76022
1706	†	**CX**	A	*SC*	BI	76094	62035	70713	76040
1707	†	**CX**	A	*SC*	BI	76084	62025	70703	76030
1708	†	**N**	A	*SC*	BI	76110	62051	70729	76056
1709	†	**N**	A	*SC*	BI	76103	62044	70722	76049
1710	†	**CX**	A	*SC*	BI	76078	62019	70697	76024
1711	‡	**N**	A	*SC*	BI	76114	62055	71766	76060
1712	†	**N**	A	*SC*	BI	76079	62020	70698	76025
1713	†	**N**	A	*SC*	BI	76128	62069	71767	76074
1714	†	**N**	A	*SC*	BI	76077	62018	70696	76023
1717	†	**N**	A	*SC*	BI	76083	62024	70702	76029
1719	†	**CX**	A	*SC*	BI	76116	62057	70719	76062
1720	†	**N**	A	*SC*	BI	76098	62039	71769	76044
1721	†	**CX**	A	*SC*	BI	76090	62031	70709	76036
1722	‡	**CX**	A	*SC*	BI	76106	62047	70725	76052
1724	†	**CX**	A	*SC*	BI	76120	62061	71770	76066
1725	†	**CX**	A	*SC*	BI	76088	62029	70707	76034
1726	†	**CX**	A	*SC*	BI	76109	62050	70728	76055
1727	†	**CX**	A	*SC*	BI	76111	62052	70730	76057
1731	†	**CX**	A	*SC*	BI	76095	62036	70714	76041
1733	†	**CX**	A	*SC*	BI	76122	62063	71047	76068
1734	†	**CX**	A	*SC*	BI	76063	62054	71044	76059
1735	†	**CX**	A	*SC*	BI	76117	62058	71050	76051
1736	†	**U**	A	*SC*	BI	76124	62065	71052	76070
1737	†	**U**	A	*SC*	BI	76121	62062	71058	76067
1738	†	**CX**	A	*SC*	BI	76129	62064	71046	76069
1739	†	**CX**	A	*SC*	BI	76123	62070	71066	76075
1740	†	**CX**	A	*SC*	BI	76126	62067	71097	76072
1741	†	**CX**	A	*SC*	BI	76089	62030	70708	76035
1742		**U**	A	*SE*	RM	76086	62027	70705	76032
1743	†	**CX**	A	*SC*	BI	76118	62059	71065	76064
1744	†	**CX**	A	*SC*	BI	76127	62068	71064	76073
1745	†	**CX**	A	*SC*	BI	76085	62026	70704	76031
1746	‡	**CX**	A	*SC*	BI	76091	62032	70710	76037
1747	†	**CX**	A	*SC*	BI	76093	62034	70712	76026
1748		**U**	A	*SE*	RM	76115	62056	71067	76061
1750	†	**CX**	A	*SC*	BI	76080	62021	70699	76039
1751	†	**CX**	A	*SC*	BI	76125	62066	71051	76071
1752	†	**N**	A	*SC*	BI	76119	62060	70717	76065
1753	†	**CX**	A	*SC*	BI	76102	62043	70721	76048
Spare		**N**	A		AF(S)		62053	71068	76058

Class 421/4. DTC(A)/(†‡DTS)–MBS–TS–DTC(B). Mk. 6 motor bogies.

1801	†	**N**	P	*SC*	BI	76848	71095	62415	76777
1802	†	**N**	P	*SC*	BI	76754	62392	71072	76825
1803	†	**N**	A	*SC*	BI	76780	62418	71098	76851
1804	†	**N**	A	*SC*	BI	76778	62416	71096	76849
1805	†	**N**	A	*SC*	BI	76782	62420	71100	76853
1806	*	**N**	H	*SE*	RM	76783	62421	71101	76854
1807	*	**N**	H	*SE*	RM	76784	62422	71102	76855
1808	*	**N**	H	*SE*	RM	76785	62423	71103	76856
1809	*	**N**	H	*SE*	RM	76786	62424	71104	76857
1810	*	**N**	H	*SE*	RM	76787	62425	71105	76858
1811	*	**N**	H	*SE*	RM	76781	62419	71099	76852
1812	*d	**N**	H	*SE*	RM	76757	62395	71075	76828
1813	*	**N**	H	*SE*	RM	76859	62430	71106	76860
1831	†	**CX**	A	*SC*	BI	76598	62304	70984	76628
1832	†	**CX**	A	*SC*	BI	76719	62357	71037	76790
1833	†	**CX**	A	*SC*	BI	76582	62288	70968	76612
1834	†	**CX**	A	*SC*	BI	76566	62282	70988	76576
1835	†	**CX**	A	*SC*	BI	76601	62307	70987	76631
1837	†	**CX**	A	*SC*	BI	76722	62360	71040	76793
1839	*	**N**	H	*SE*	RM	76607	62313	70993	76637
1840	*	**N**	H	*SE*	RM	76724	62362	71042	76795
1841	*	**N**	H	*SE*	RM	76603	62309	70989	76633
1842	*	**N**	H	*SE*	RM	76725	62363	71043	76796
1843	*	**N**	H	*SE*	RM	76731	62369	71049	76802
1845	†	**CX**	A	*SC*	BI	76599	62305	70985	76718
1846	†	**CX**	A	*SC*	BI	76737	62375	71055	76808
1847	†	**CX**	A	*SC*	BI	76600	62306	70986	76630
1848	†	**CX**	A	*SC*	BI	76605	62311	70991	76635
1850	†	**CX**	A	*SC*	BI	76629	62356	71036	76789
1851	†	**CX**	A	*SC*	BI	76721	62359	71039	76792
1853	†	**CX**	A	*SC*	BI	76606	62312	70992	76636
1854	†	**CX**	A	*SC*	BI	76738	62376	71056	76809
1855	†	**CX**	A	*SC*	BI	76720	62358	71038	76791
1856	†	**CX**	A	*SC*	BI	76739	62377	71057	76810
1857	†	**CX**	A	*SC*	BI	76610	62316	70996	76640
1858	‡	**CX**	A	*SC*	BI	76604	62310	70990	76634
1859	‡	**CX**	A	*SC*	BI	76727	62365	71045	76798
1860	‡	**CX**	A	*SC*	BI	76752	62390	71070	76823
1861	‡	**CX**	A	*SC*	BI	76735	62373	71053	76806
1862	†	**CX**	A	*SC*	BI	76736	62374	71054	76807
1863	†	**CX**	A	*SC*	BI	76742	62380	71060	76813
1864	†	**CX**	A	*SC*	BI	76741	62379	71059	76812
1865	†	**CX**	A	*SC*	BI	76745	62383	71063	76639
1866	†	**CX**	A	*SC*	BI	76743	62381	71061	76814
1867	†	**CX**	A	*SC*	BI	76744	62382	71062	76815
1868	†	**CX**	A	*SC*	BI	76751	62389	71069	76822
1869	†	**CX**	A	*SC*	BI	76753	62391	71071	76804
1870	*	**CX**	H	*SE*	RM	76108	62409	71089	76842
1871	*	**N**	H	*SE*	RM	76756	62394	71074	76827

1872	*	**N**	H	*SE*	RM	76771	62396	71076	76829
1873	*	**N**	H	*SE*	RM	76759	62397	71077	76830
1874	†	**CX**	A	*SC*	BI	76755	62393	71073	76826
1876	*	**N**	H	*SE*	RM	76761	62399	71079	76832
1877	*	**N**	H	*SE*	RM	76763	62401	71081	76834
1878	*	**N**	H	*SE*	RM	76768	62406	71086	76839
1879	*	**N**	H	*SE*	RM	76760	62398	71078	76831
1880		**ST**	H	*SW*	FR	76770	62408	71088	76841
1881		**ST**	H	*SW*	FR	76762	62400	71080	76833
1882		**ST**	H	*SW*	FR	76765	62403	71083	76836
1883		**ST**	H	*SW*	FR	76764	62402	71082	76835
1884		**N**	H	*SW*	FR	76767	62405	71085	76838
1885		**N**	H	*SW*	FR	76769	62407	71087	76840
1886		**N**	H	*SW*	FR	76772	62410	71090	76843
1887		**N**	H	*SW*	FR	76766	62404	71084	76837
1888		**N**	H	*SW*	FR	76773	62411	71091	76844
1889		**N**	H	*SW*	FR	76774	62412	71092	76845
1890		**N**	H	*SW*	FR	76775	62413	71093	76846
1891		**N**	H	*SW*	FR	76776	62414	71094	76847
Spare	•	**BG**	A		ZG(S)			70995	

Class 421/6. DTS–MBS–TS–DTC(B). Mk. 6 motor bogies.

1901	†	**CX**	P	*SC*	BI	76082	62023	70701	76028
1902	†	**CX**	P	*SC*	BI	76100	62041	71768	76046
1903	†	**CX**	A	*SC*	BI	76081	62022	70700	76027
1904	†	**CX**	A	*SC*	BI	76107	62048	70726	76053
1905	‡	**CX**	A	*SC*	BI	76099	62040	70718	76045
1906	†	**CX**	A	*SC*	BI	76105	62046	70724	76113
1907	†	**CX**	A	*SC*	BI	76104	62045	70723	76050
1908	†	**CX**	A	*SC*	BI	76096	62037	70715	76042

CLASS 422 Spare Buffet Car

For details see pages 56–57.

Spare TRBS.

Spare	**N**	P	ZG(S)	69302	69304	69306	69307
Spare	**N**	P	ZG(S)	69310	69316	69318	69333
Spare	**N**	P	ZG(S)	69335			

CLASS 412 4-Bep Express Unit

For details see pages 54–55.

Class 412. DMS(A)–TBC–TRBS–DMS(B).

2301	†	**ST**	P	*SW*	FR	61804	70607	69341	61805
2302	†	**ST**	P	*SW*	FR	61774	70592	69342	61809
2303	†	**ST**	P	*SW*	FR	61954	70656	69343	61955
2304	†	**ST**	P	*SW*	FR	61736	70573	69344	61737
2305	†	**ST**	P	*SW*	FR	61798	70354	69345	61799
2306	†	**ST**	P	*SW*	FR	61808	70609	69346	61775
2307	†	**ST**	P	*SW*	FR	61802	70606	69347	61803

CLASS 442 5-Wes Express Unit

DTF–TS–MBLS–TSW–DTS. Gangwayed throughout. Air conditioned.
Traction Motors: Four English Electric 546 of 300 kW each.
Dimensions: 23.15 (DTF & DTS) or 23.00 (MBLS, TS & TSW) x 2.74 x 3.81m.
Doors: Power operated sliding plug.
Maximum Speed: 100 mph. **Bogies:** Mk. 6/T4.
Couplings: Buckeye. **Multiple Working:** SR type.

62937–62960. MBLS. Dia. ED268. Lot No. 31034 BREL Derby 1988–89. Modified Adtranz Crewe 1998. –/30 1W. 55.2 t.
71818–71841. TS. Dia. EH288. Lot No. 31032 BREL Derby 1988–89. –/80 2T. 35.3 t.
71842–71865. TSW. Dia. EH289. Lot No. 31033 BREL Derby 1988–89. –/76 1W 2T. 35.4 t.
77382–77405. DTF. Dia. EE160. Lot No. 31030 BREL Derby 1988–89. 50/– 1T. 39.1 t.
77406–77429. DTS. Dia. EE273. Lot No. 31031 BREL Derby 1988–89. –/78 1T. 39.1 t.

2401	**SW**	A	*SW*	BM	77382	71818	62937	71842	77406
2402	**SW**	A	*SW*	BM	77383	71819	62938	71843	77407
2403	**SW**	A	*SW*	BM	77384	71820	62941	71844	77408
2404	**SW**	A	*SW*	BM	77385	71821	62939	71845	77409
2405	**SW**	A	*SW*	BM	77386	71822	62944	71846	77410
2406	**SW**	A	*SW*	BM	77389	71823	62942	71847	77411
2407	**SW**	A	*SW*	BM	77388	71824	62943	71848	77412
2408	**SW**	A	*SW*	BM	77387	71825	62945	71849	77413
2409	**SW**	A	*SW*	BM	77390	71826	62946	71850	77414
2410	**SW**	A	*SW*	BM	77391	71827	62948	71851	77415
2411	**SW**	A	*SW*	BM	77392	71828	62940	71858	77422
2412	**SW**	A	*SW*	BM	77393	71829	62947	71853	77417
2413	**SW**	A	*SW*	BM	77394	71830	62949	71854	77418
2414	**SW**	A	*SW*	BM	77395	71831	62950	71855	77419
2415	**SW**	A	*SW*	BM	77396	71832	62951	71856	77420
2416	**SW**	A	*SW*	BM	77397	71833	62952	71857	77421
2417	**SW**	A	*SW*	BM	77398	71834	62953	71852	77416
2418	**SW**	A	*SW*	BM	77399	71835	62954	71859	77423

2419	**SW**	A	*SW*	BM	77400 71836 62955 71860 77424
2420	**SW**	A	*SW*	BM	77401 71837 62956 71861 77425
2421	**SW**	A	*SW*	BM	77402 71838 62957 71862 77426
2422	**SW**	A	*SW*	BM	77403 71839 62958 71863 77427
2423	**SW**	A	*SW*	BM	77404 71840 62959 71864 77428
2424	**SW**	A	*SW*	BM	77405 71841 62960 71865 77429

Names (carried on MBLS):

2401	BEAULIEU
2402	COUNTY OF HAMPSHIRE
2403	THE NEW FOREST
2404	BOROUGH OF WOKING
2405	CITY OF PORTSMOUTH
2406	VICTORY
2407	THOMAS HARDY
2408	COUNTY OF DORSET
2420	CITY SOUTHAMPTON
2422	OPERATION OVERLORD
2409	BOURNEMOUTH ORCHESTRAS
2410	MERIDIAN TONIGHT
2412	SPECIAL OLYMPICS
2415	MARY ROSE
2416	MUM IN A MILLION 1997 – DOREEN SCANLON
2418	WESSEX CANCER TRUST
2419	BBC SOUTH TODAY
2423	COUNTY OF SURREY

CLASS 423 — 4-Vep or 4-Vop Unit

Various formations, see below. Gangwayed throughout.
Traction Motors: Four English Electric 507 of 185 kW each.
Dimensions: 20.18 x 2.82 x 3.84 m. **Doors**: Manually operated slam.
Maximum Speed: 90 mph. **Multiple Working**: SR type.
Couplings: Buckeye. **Bogies**: Mk. 4/B5 (SR).

62121–62140. MBS. Dia. ED266. Lot No. 30760 Derby 1967. –/76. 49.0 t.
62182–62216. MBS. Dia. ED266. Lot No. 30773 York 1967–68. –/76. 49.0 t.
62217–62266. MBS. Dia. ED266. Lot No. 30794 York 1968–69. –/76. 49.0 t.
62267–62276. MBS. Dia. ED266. Lot No. 30800 York 1970. –/76. 49.0 t.
62317–62354. MBS. Dia. ED266. Lot No. 30813 York 1970–73. –/76. 49.0 t.
62435–62475. MBS. Dia. ED266. Lot No. 30851 York 1973–74. –/76. 49.0 t.
70781–70800. TS. Dia. EH291. Lot No. 30759 Derby 1967. –/98. 31.5 t.
70872–70906. TS. Dia. EH291. Lot No. 30772 York 1967–68. –/98. 31.5 t.
70907–70956. TS. Dia. EH291. Lot No. 30793 York 1968–69. –/98. 31.5 t.
70957–70966. TS. Dia. EH291. Lot No. 30801 York 1970. –/98. 31.5 t.
70997–71034. TS. Dia. EH291. Lot No. 30812 York 1970–73. –/98. 31.5 t.
71115–71155. TS. Dia. EH291. Lot No. 30852 York 1973–74. –/98. 31.5 t.
76230–76269. DTC. Dia. EE373. Lot No. 30758 York 1967. 18/46 (†§ 12/52)1T. 35.0 t.
76275. DTS. Dia. EE266. Built as loco-hauled vehicle to Lot No. 30086 Eastleigh 1953–55. Converted to Lot No. 30764 York 1966. –/64. 32.0 t.
76333–76402. DTC (*‡ DTS). Dia. EE373 (* EE281; ‡ EE278). Lot No. 30771 York 1967–68. (*Converted Adtranz, Chart Leacon 1999–2000). 18/46 († 12/52)1T ; (* –/70; ‡ –/88 0T). 32.5 t.
76441–76540. DTC. Dia. EE373. Lot No. 30792 York 1968–69. 18/46 (†‡ 12/52) 1T. 32.5 t.
76541–76560. DTC. Dia. EE373. Lot No. 30799 York 1970. 18/46 († 12/52) 1T. 32.5 t.
76641–76716. DTC. Dia. EE373. Lot No. 30811 York 1970–73. 18/46 († 12/52) 1T. 32.5 t.
76861–76942. DTC. Dia. EE368. Lot No. 30853 York 1973–74. 18/46 († 12/52) 1T. 32.5 t.

Class 423/1. 4-Vep. DTC–MBS–TS–DTC.

3401		**ST**	H	*SW*	WD	76230	62276	70781	76231
3402		**ST**	H	*SW*	WD	76233	62123	70782	76232
3403		**CX**	H	*SC*	BI	76234	62254	70783	76235
3404		**ST**	H	*SW*	WD	76378	62261	70894	76236
3405		**ST**	H	*SW*	WD	76239	62271	70785	76238
3406		**ST**	H	*SW*	WD	76241	62130	70786	76240
3407		**ST**	H	*SW*	WD	76243	62348	70787	76242
3408		**ST**	H	*SW*	WD	76244	62435	70788	76245
3409		**ST**	H	*SW*	WD	76246	62239	70789	76247
3410		**ST**	H	*SW*	WD	76369	62442	70790	76249
3411		**ST**	H	*SW*	WD	76250	62342	70791	76251
3412	†	**N**	A	*SE*	RM	76252	62340	70792	76253
3413		**ST**	H	*SW*	WD	76255	62441	70793	76254
3414		**ST**	H	*SW*	WD	76257	62446	70794	76248
3415		**N**	H	*SW*	WD	76258	62462	70795	76259
3416	†	**N**	A	*SE*	RM	76261	62451	70796	76260
3417		**ST**	H	*SW*	WD	76262	62236	70797	76263
3418		**ST**	H	*SW*	WD	76265	62133	70875	76264
3419		**ST**	H	*SW*	WD	76267	62354	70799	76266
3420		**ST**	H	*SW*	WD	76269	62349	70800	76268
3421	†	**CX**	A	*SE*	RM	76889	62449	71129	76890
3422	†	**CX**	A	*SE*	RM	76372	62201	70891	76371
3423	†	**CX**	A	*SE*	RM	76452	62222	70912	76451
3424	†	**CX**	A	*SE*	RM	76354	62185	70882	76353
3425		**ST**	H	*SW*	WD	76338	62192	70874	76358
3426		**ST**	H	*SW*	WD	76386	62208	70898	76385
3427		**ST**	H	*SW*	WD	76374	62184	70892	76373
3428		**ST**	H	*SW*	WD	76454	62223	70913	76453
3429		**ST**	H	*SW*	WD	76334	62202	70872	76333
3430		**ST**	H	*SW*	WD	76348	62189	70879	76347
3431		**ST**	H	*SW*	WD	76458	62182	70915	76457
3432		**ST**	H	*SW*	WD	76400	62225	70905	76399
3433		**ST**	H	*SW*	WD	76444	62215	70908	76443
3434		**ST**	H	*SW*	WD	76462	62218	70917	76461
3435		**CX**	P	*SC*	BI	76342	62228	70876	76341
3436		**CX**	P	*SC*	BI	76350	62190	70880	76349
3437		**CX**	P	*SC*	BI	76346	62186	70878	76345
3445	†	**CX**	A	*SE*	RM	76450	62242	70911	76449
3446	†	**CX**	A	*SE*	RM	76532	62243	70952	76531
3447	†	**CX**	A	*SE*	RM	76380	62199	70895	76379
3448	†	**CX**	A	*SE*	RM	76376	62221	70886	76375
3449	†	**CX**	A	*SE*	RM	76336	62205	70873	76335
3450	†	**CX**	A	*SE*	RM	76460	62203	70916	76459
3451	†	**N**	A	*SE*	RM	76488	62240	70930	76487
3452	†	**CX**	A	*SE*	RM	76340	62183	71021	76690
3453	†	**CX**	A	*SE*	RM	76382	62226	70896	76381
3454	†	**CX**	A	*SE*	RM	76390	62200	70798	76389
3455		**ST**	H	*SW*	WD	76388	62206	70899	76387
3456		**ST**	H	*SW*	WD	76456	62210	70914	76455

3457		**ST**	H	*SW*	WD	76392	62197	70901	76391
3458		**ST**	H	*SW*	WD	76394	62209	70902	76393
3459		**ST**	H	*SW*	WD	76396	62224	70903	76395
3466		**ST**	H	*SW*	WD	76464	62214	70918	76463
3467		**ST**	H	*SW*	WD	76446	62217	70909	76445
3468		**ST**	H	*SW*	WD	76448	62267	70910	76447
3469		**N**	H	*SW*	WD	76546	62219	70959	76545
3470		**N**	H	*SW*	WD	76496	62220	70934	76495
3471	†	**CX**	A	*SE*	RM	76498	62269	70935	76497
3472	†	**N**	A	*SE*	RM	76500	62244	70936	76499
3473	‡	**CX**	A	*SE*	RM	76502	62245	70937	76339
3474	†	**N**	A	*SE*	RM	76504	62246	70938	76503
3475	†	**N**	A	*SE*	RM	76552	62270	70962	76551
3479		**CX**	H	*SC*	BI	76655	62272	71004	76656
3480		**ST**	H	*SC*	WD	76474	62323	70923	76473
3481		**ST**	H	*SW*	WD	76647	62324	70900	76648
3482		**CX**	H	*SC*	BI	76657	62320	71005	76658
3483		**CX**	H	*SC*	BI	76661	62233	71007	76662
3484		**CX**	H	*SC*	BI	76476	62325	70924	76475
3485		**CX**	H	*SC*	BI	76508	62327	70940	76507
3486		**CX**	H	*SC*	BI	76478	62234	70925	76477
3487	†	**N**	A	*SE*	RM	76645	62250	70941	76509
3488		**CX**	H	*SC*	BI	76663	62235	71008	76664
3489		**CX**	H	*SC*	BI	76665	62251	71009	76666
3490		**CX**	H	*SC*	BI	76695	62328	71024	76696
3491	†	**N**	A	*SE*	RM	76337	62436	70927	76481
3492	†	**N**	A	*SE*	RM	76667	62344	71010	76668
3493	†	**N**	A	*SE*	RM	76669	62237	71011	76670
3494	†	**N**	A	*SE*	RM	76675	62330	71014	76676
3495	†	**N**	A	*SE*	RM	76699	62331	71026	76700
3496	†	**N**	A	*SE*	RM	76673	62334	71013	76674
3497	†	**N**	A	*SE*	RM	76671	62346	71012	76672
3498	†	**N**	A	*SE*	RM	76701	62333	71027	76702
3499	†	**N**	A	*SE*	RM	76901	62347	71135	76902
3500	†	**N**	A	*SE*	RM	76470	62455	70921	76469
3501		**CX**	P	*SC*	BI	76512	62332	70942	76511
3503		**CX**	P	*SC*	BI	76681	62231	71017	76682
3504		**CX**	P	*SC*	BI	76711	62351	71032	76712
3505		**CX**	P	*SC*	BI	76472	62352	70922	76471
3506		**N**	P	*SC*	BI	76554	62317	70963	76553
3507		**N**	P	*SC*	BI	76558	62232	70965	76557
3508		**ST**	H	*SW*	WD	76643	62273	70998	76644
3509		**ST**	H	*SW*	WD	76560	62275	70966	76559
3510		**ST**	H	*SW*	WD	76641	62318	70997	76642
3511	†	**CX**	A	*SE*	RM	76893	62135	70999	76646
3512		**CX**	P	*SC*	BI	76679	62337	71016	76680
3514		**CX**	P	*SC*	BI	76683	62136	71018	76684
3515		**CX**	P	*SC*	BI	76544	62319	70958	76543
3516		**ST**	H	*SW*	WD	76693	62268	71023	76694
3517		**CX**	P	*SC*	BI	76685	62338	71019	76686
3518		**CX**	P	*SC*	BI	76689	62343	70887	76363

3519		**ST**	H	*SW*	WD	76556	62274	70964	76555
3520		**ST**	H	*SW*	WD	76697	62131	71025	76698
3521	†	**N**	A	*SE*	RM	76484	62345	70928	76483
3523		**CX**	H	*SC*	BI	76651	62139	71002	76652
3524		**CX**	H	*SC*	BI	76466	62322	70919	76370
3526		**CX**	P	*SC*	BI	76524	62255	70948	76523
3527		**CX**	P	*SC*	BI	76520	62326	70946	76519
3528		**CX**	P	*SC*	BI	76518	62258	70945	76517
3529		**CX**	H	*SC*	BI	76659	62257	71006	76660
3530		**CX**	H	*SC*	BI	76468	62256	70920	76467
3531		**CX**	H	*SC*	BI	76649	62230	71001	76650
3532		**CX**	P	*SC*	BI	76528	62321	70950	76527
3535		**CX**	P	*SC*	BI	76677	62335	71015	76678
3536		**N**	H	*SW*	WD	76384	62207	70897	76383
3539		**N**	H	*SW*	WD	76861	62122	71115	76862
3540		**N**	H	*SW*	WD	76863	62128	71116	76864
3542		**ST**	H	*SW*	WD	76480	62127	70926	76479
3543	†	**N**	A	*SE*	RM	76899	62137	71134	76900
3544	†	**N**	A	*SE*	RM	76892	62454	71131	76894
3545	†	**N**	A	*SE*	RM	76875	62121	71122	76876
3546		**CX**	P	*SC*	BI	76687	62339	71020	76688
3547	†	**N**	A	*SE*	RM	76895	62126	71132	76896
3548	†	**N**	A	*SE*	RM	76903	62452	71136	76904
3549		**CX**	P	*SC*	BI	76707	62132	71030	76708
3551		**CX**	P	*SC*	BI	76465	62456	71033	76714
3552		**ST**	H	*SW*	WD	76715	62353	71034	76716
3553	†	**N**	A	*SE*	RM	76913	62241	71141	76914
3554	†	**CX**	A	*SE*	RM	76905	62461	71137	76906
3555		**ST**	H	*SW*	WD	76865	62140	71117	76866
3556	†	**CX**	A	*SE*	RM	76885	62457	71127	76886
3557		**ST**	H	*SW*	WD	76869	62437	71119	76870
3558		**ST**	H	*SW*	WD	76352	62447	70881	76351
3559		**ST**	H	*SW*	WD	76486	62439	70929	76485
3560	†	**CX**	A	*SE*	RM	76897	62191	71133	76898
3561		**ST**	H	*SW*	WD	76867	62453	71118	76868
3562	†	**CX**	A	*SE*	RM	76907	62129	71138	76908
3563		**ST**	H	*SW*	WD	76873	62438	71121	76874
3564	†	**CX**	A	*SE*	RM	76883	62458	71126	76884
3565	†	**CX**	A	*SE*	RM	76877	62134	71123	76878
3566	†	**CX**	A	*SE*	RM	76915	62443	71142	76916
3567		**ST**	H	*SW*	WD	76871	62138	71120	76872
3568	†	**CX**	A	*SE*	RM	76887	62440	71128	76888
3569		**ST**	H	*SW*	WD	76344	62448	70877	76343
3570	†	**CX**	A	*SE*	RM	76909	62187	71139	76910
3571	†	**CX**	A	*SE*	RM	76927	62463	71148	76928
3572	†	**CX**	A	*SE*	RM	76879	62468	71124	76880
3573	†	**CX**	A	*SE*	RM	76919	62444	71144	76920
3574	†	**CX**	A	*SE*	RM	76929	62464	71149	76930
3575	†	**CX**	A	*SE*	RM	76931	62469	71150	76932
3576		**ST**	H	*SW*	WD	76362	62196	70890	76361
3577	†	**CX**	A	*SE*	RM	76933	62459	71151	76934

3578		**ST**	H	*SW*	WD	76356	62193	70883	76355
3579	†	**CX**	A	*SE*	RM	76935	62471	71152	76936
3580		**ST**	H	*SW*	WD	76360	62195	70885	76359
3581		**ST**	H	*SW*	WD	76366	62198	70888	76365
3582	†	**CX**	A	*SE*	RM	76891	62472	71130	76275
3583	†	**CX**	A	*SE*	RM	76937	62450	71153	76938
3584	†	**CX**	A	*SE*	RM	76881	62473	71125	76882
3585	†	**CX**	A	*SE*	RM	76939	62445	71154	76940
3586	†	**CX**	A	*SE*	RM	76921	62474	71145	76922
3587	†	**CX**	A	*SE*	RM	76925	62465	71147	76926
3588	†	**CX**	A	*SE*	RM	76923	62467	71146	76924
3589	†	**CX**	A	*SE*	RM	76911	62466	71140	76912
3590	†	**CX**	A	*SE*	RM	76941	62460	71155	76942
3591	†	**CX**	A	*SE*	RM	76917	62475	71143	76918
3801	†	**CX**	P	*SE*	RM	76522	62229	70947	76521
3802	†	**CX**	P	*SE*	RM	76534	62188	70953	76533
3803	†	**CX**	P	*SE*	RM	76494	62263	70933	76493
3804	†	**CX**	P	*SE*	RM	76368	62204	70889	76367
3805	†	**CX**	P	*SE*	RM	76540	62211	70956	76539
3806	†	**N**	P	*SE*	RM	76538	62212	70955	76537
3807	†	**CX**	P	*SE*	RM	76542	62264	70957	76541
3808	†	**N**	P	*SE*	RM	76550	62248	70961	76549
3809		**N**	P	*SW*	WD	76516	62253	70944	76515
3810		**N**	P	*SW*	WD	76709	62252	71031	76710
3811		**N**	P	*SW*	WD	76514	62249	70943	76513
3812		**ST**	P	*SW*	WD	76703	62238	71028	76704
Spare		**N**	H		WD(S)		62470		

Class 423/2. 4-Vop. DTS–MBS–TS–DTS. Declassified units for Connex South Central 'South London Metro' services.

No.	*Former No.*									
3901	3439	*	**CX**	P	*SC*	BI	76402	62227	70906	76401
3902	3533	*	**CX**	P	*SC*	BI	76364	62260	70949	76525
3903	3462	*	**CX**	P	*SC*	BI	76536	62213	70954	76535
3904	3513	*	**CX**	P	*SC*	BI	76691	62336	71022	76692
3905	3463	*	**CX**	P	*SC*	BI	76398	62266	70904	76397
3906	3550	*	**CX**	P	*SC*	BI	76490	62350	70931	76489
3907	3534	*	**CX**	P	*SC*	BI	76506	62259	70939	76505
3908	3464	*	**CX**	P	*SC*	BI	76442	62265	70907	76441
3909	3522	*	**CX**	P	*SC*	BI	76705	62341	71029	76706
3910	3438	*	**CX**	P	*SC*	BI	76530	62262	70951	76529
3911	3476	*	**CX**	P	*SC*	BI	76548	62247	70960	76547
3912	3442	*	**CX**	P	*SC*	BI	76492	62216	70932	76491
3913	3478	*	**CX**	P	*SC*	BI	76653	62125	71003	76654
3914										
3915										
3916										
3917										
3918										
3919										

CLASS 455 4-Car Unit

DTS–MS–TS–DTS. Gangwayed throughout. Disc brakes (5913–15 have tread brakes).
Traction Motors: Four GEC507-20J of 185 kW each.
Dimensions: 19.92 x 2.82 x 3.77 (3.58 Class 455/7 TSO) m.
Maximum Speed: 75 mph. **Couplers**: Tightlock.
Doors: Power operated sliding.
Bogies: BP27/BT13 (Classes 455/8 and 455/9) or BX1 (Class 455/7).
Multiple Working: Classes 455–457.

Class 455/7. Built as 3-car units, augmented with ex-Class 508 TSO. Pressure heating & ventilation.

62783–62825. MS. Dia. EC203. Lot No. 30975 BREL York 1984–85. –/84. 45.0 t.
71526–71568. TS. Dia. EH219. Lot No. 30944 BREL York 1977–80. –/86. 25.5 t.
77727–77812. DTS. Dia. EE218. Lot No. 30976 BREL York 1984–85. –/74. 29.5 t.

5701	**N**	P	*SW*	WD	77727	62783	71545	77728
5702	**N**	P	*SW*	WD	77729	62784	71547	77730
5703	**N**	P	*SW*	WD	77731	62785	71540	77732
5704	**N**	P	*SW*	WD	77733	62786	71548	77734
5705	**N**	P	*SW*	WD	77735	62787	71565	77736
5706	**N**	P	*SW*	WD	77737	62788	71534	77738
5707	**N**	P	*SW*	WD	77739	62789	71536	77740
5708	**N**	P	*SW*	WD	77741	62790	71560	77742
5709	**N**	P	*SW*	WD	77743	62791	71532	77744
5710	**N**	P	*SW*	WD	77745	62792	71566	77746
5711	**N**	P	*SW*	WD	77747	62793	71542	77748
5712	**N**	P	*SW*	WD	77749	62794	71546	77750
5713	**N**	P	*SW*	WD	77751	62795	71567	77752
5714	**ST**	P	*SW*	WD	77753	62796	71539	77754
5715	**ST**	P	*SW*	WD	77755	62797	71535	77756
5716	**ST**	P	*SW*	WD	77757	62798	71564	77758
5717	**N**	P	*SW*	WD	77759	62799	71528	77760
5718	**N**	P	*SW*	WD	77761	62800	71557	77762
5719	**ST**	P	*SW*	WD	77763	62801	71558	77764
5720	**ST**	P	*SW*	WD	77765	62802	71568	77766
5721	**ST**	P	*SW*	WD	77767	62803	71553	77768
5722	**ST**	P	*SW*	WD	77769	62804	71533	77770
5723	**ST**	P	*SW*	WD	77771	62805	71526	77772
5724	**ST**	P	*SW*	WD	77773	62806	71561	77774
5725	**ST**	P	*SW*	WD	77775	62807	71541	77776
5726	**ST**	P	*SW*	WD	77777	62808	71556	77778
5727	**ST**	P	*SW*	WD	77779	62809	71562	77780
5728	**ST**	P	*SW*	WD	77781	62810	71527	77782
5729	**ST**	P	*SW*	WD	77783	62811	71550	77784
5730	**ST**	P	*SW*	WD	77785	62812	71551	77786
5731	**ST**	P	*SW*	WD	77787	62813	71555	77788
5732	**ST**	P	*SW*	WD	77789	62814	71552	77790
5733	**ST**	P	*SW*	WD	77791	62815	71549	77792

5734	**ST**	P	*SW*	WD	77793	62816	71531	77794
5735	**ST**	P	*SW*	WD	77795	62817	71563	77796
5736	**ST**	P	*SW*	WD	77797	62818	71554	77798
5737	**ST**	P	*SW*	WD	77799	62819	71544	77800
5738	**ST**	P	*SW*	WD	77801	62820	71529	77802
5739	**ST**	P	*SW*	WD	77803	62821	71537	77804
5740	**ST**	P	*SW*	WD	77805	62822	71530	77806
5741	**ST**	P	*SW*	WD	77807	62823	71559	77808
5742	**ST**	P	*SW*	WD	77809	62824	71543	77810
5750	**ST**	P	*SW*	WD	77811	62825	71538	77812

Names:

5711 SPIRIT OF RUGBY
5731 VARIETY CLUB
5735 The Royal Borough of Kingston
5750 Wimbledon Train Care

Class 455/8. Pressure heating & ventilation.

62709–62782. MS. Dia. EC203. Lot No. 30973 BREL York 1982–84. –/84. 45.6 t.
71637–71710. TS. Dia. EH221. Lot No. 30974 BREL York 1982–84. –/84. 27.1 t.
77579–77726. DTS. Dia. EE218. Lot No. 30972 BREL York 1982–84. –/74. 29.5 t.

5801	**N**	H	*SC*	SU	77579	62709	71637	77580
5802	**N**	H	*SC*	SU	77581	62710	71664	77582
5803	**N**	H	*SC*	SU	77583	62711	71639	77584
5804	**CX**	H	*SC*	SU	77585	62712	71640	77586
5805	**N**	H	*SC*	SU	77587	62713	71641	77588
5806	**N**	H	*SC*	SU	77589	62714	71642	77590
5807	**N**	H	*SC*	SU	77591	62715	71643	77592
5808	**N**	H	*SC*	SU	77593	62716	71644	77594
5809	**N**	H	*SC*	SU	77595	62717	71645	77596
5810	**N**	H	*SC*	SU	77597	62718	71646	77598
5811	**N**	H	*SC*	SU	77599	62719	71647	77600
5812	**N**	H	*SC*	SU	77601	62720	71648	77602
5813	**N**	H	*SC*	SU	77603	62721	71649	77604
5814	**N**	H	*SC*	SU	77605	62722	71650	77606
5815	**N**	H	*SC*	SU	77607	62723	71651	77608
5816	**N**	H	*SC*	SU	77609	62724	71652	77633
5817	**N**	H	*SC*	SU	77611	62725	71653	77612
5818	**N**	H	*SC*	SU	77613	62726	71654	77614
5819	**N**	H	*SC*	SU	77615	62727	71655	77616
5820	**N**	H	*SC*	SU	77617	62728	71656	77618
5821	**N**	H	*SC*	SU	77619	62729	71657	77620
5822	**N**	H	*SC*	SU	77621	62730	71658	77622
5823	**N**	H	*SC*	SU	77623	62731	71659	77624
5824	**N**	H	*SC*	SU	77637	62732	71660	77626
5825	**N**	H	*SC*	SU	77627	62733	71661	77628
5826	**N**	H	*SC*	SU	77629	62734	71662	77630
5827	**N**	H	*SC*	SU	77610	62735	71663	77632
5828	**N**	H	*SC*	SU	77631	62736	71638	77634
5829	**N**	H	*SC*	SU	77635	62737	71665	77636

5830	**N**	H	*SC*	SU	77625	62743	71666	77638
5831	**N**	H	*SC*	SU	77639	62739	71667	77640
5832	**N**	H	*SC*	SU	77641	62740	71668	77642
5833	**N**	H	*SC*	SU	77643	62741	71669	77644
5834	**N**	H	*SC*	SU	77645	62742	71670	77646
5835	**N**	H	*SC*	SU	77647	62738	71671	77648
5836	**N**	H	*SC*	SU	77649	62744	71672	77650
5837	**N**	H	*SC*	SU	77651	62745	71673	77652
5838	**N**	H	*SC*	SU	77653	62746	71674	77654
5839	**N**	H	*SC*	SU	77655	62747	71675	77656
5840	**N**	H	*SC*	SU	77657	62748	71676	77658
5841	**N**	H	*SC*	SU	77659	62749	71677	77660
5842	**N**	H	*SC*	SU	77661	62750	71678	77662
5843	**N**	H	*SC*	SU	77663	62751	71679	77664
5844	**N**	H	*SC*	SU	77665	62752	71680	77666
5845	**N**	H	*SC*	SU	77667	62753	71681	77668
5846	**N**	H	*SC*	SU	77669	62754	71682	77670
5847	**N**	P	*SW*	WD	77671	62755	71683	77672
5848	**N**	P	*SW*	WD	77673	62756	71684	77674
5849	**N**	P	*SW*	WD	77675	62757	71685	77676
5850	**N**	P	*SW*	WD	77677	62758	71686	77678
5851	**N**	P	*SW*	WD	77679	62759	71687	77680
5852	**N**	P	*SW*	WD	77681	62760	71688	77682
5853	**N**	P	*SW*	WD	77683	62761	71689	77684
5854	**N**	P	*SW*	WD	77685	62762	71690	77686
5855	**N**	P	*SW*	WD	77687	62763	71691	77688
5856	**N**	P	*SW*	WD	77689	62764	71692	77690
5857	**N**	P	*SW*	WD	77691	62765	71693	77692
5858	**N**	P	*SW*	WD	77693	62766	71694	77694
5859	**N**	P	*SW*	WD	77695	62767	71695	77696
5860	**N**	P	*SW*	WD	77697	62768	71696	77698
5861	**N**	P	*SW*	WD	77699	62769	71697	77700
5862	**N**	P	*SW*	WD	77701	62770	71698	77702
5863	**N**	P	*SW*	WD	77703	62771	71699	77704
5864	**N**	P	*SW*	WD	77705	62772	71700	77706
5865	**N**	P	*SW*	WD	77707	62773	71701	77708
5866	**N**	P	*SW*	WD	77709	62774	71702	77710
5867	**N**	P	*SW*	WD	77711	62775	71703	77712
5868	**N**	P	*SW*	WD	77713	62776	71704	77714
5869	**N**	P	*SW*	WD	77715	62777	71705	77716
5870	**N**	P	*SW*	WD	77717	62778	71706	77718
5871	**N**	P	*SW*	WD	77719	62779	71707	77720
5872	**N**	P	*SW*	WD	77721	62780	71708	77722
5873	**N**	P	*SW*	WD	77723	62781	71709	77724
5874	**N**	P	*SW*	WD	77725	62782	71710	77726

Class 455/9. Convection heating.

62826–62845. MS. Dia. EC206. Lot No. 30992 BREL York 1985. –/84. 45.6 († 48.0) t.
67400. TS. Dia. EH236. Built as a Class 210 DEMU vehicle to Lot No. 30932 BREL Derby 1981. Subsequently converted to EMU vehicle. –/84. 26.8 t.
71714–71733. TS. Dia. EH224. Lot No. 30993 BREL York 1985. –/84. 27.1 t.
77813–77852. DTS. Dia. EE226. Lot No. 30991 BREL York 1985. –/74. 29.5 t.

5901		**ST**	P	*SW*	WD	77813	62826	71714	77814
5902		**ST**	P	*SW*	WD	77815	62827	71715	77816
5903		**ST**	P	*SW*	WD	77817	62828	71716	77818
5904		**ST**	P	*SW*	WD	77819	62829	71717	77820
5905		**ST**	P	*SW*	WD	77821	62830	71725	77822
5906		**ST**	P	*SW*	WD	77823	62831	71719	77824
5907		**ST**	P	*SW*	WD	77825	62832	71720	77826
5908		**ST**	P	*SW*	WD	77827	62833	71721	77828
5909		**ST**	P	*SW*	WD	77829	62834	71722	77830
5910		**ST**	P	*SW*	WD	77831	62835	71723	77832
5911		**ST**	P	*SW*	WD	77833	62836	71724	77834
5912	†	**ST**	P		ZG(S)	77835	62837	71731	77836
5913		**ST**	P	*SW*	WD	77837	62838	71726	77838
5914		**ST**	P	*SW*	WD	77839	62839	71727	77840
5915		**ST**	P	*SW*	WD	77841	62840	71728	77842
5916		**ST**	P	*SW*	WD	77843	62841	71729	77844
5917		**ST**	P	*SW*	WD	77845	62842	71730	77846
5918		**N**	P	*SW*	WD	77847	62843	71732	77848
5919		**ST**	P	*SW*	WD	77849	62844	71718	77850
5920		**N**	P	*SW*	WD	77851	62845	71733	77852
Spare		**N**	P		WD(S)			67400	

CLASS 458 Alstom 'Juniper' 4-Jop

DMC(A)–PTS–MS–DMC(B). Gangwayed throughout. Disc and regenerative brakes. Air conditioned.
Supply System: 750 V d.c. third rail with provision for 25 kV a.c. 50 Hz overhead.
Traction Motors: Two Alstom Onix 800 of 270 kW each per power car.
Dimensions: 21.16 (DMC) or 19.94 (MS & PTS) x 2.80 x . m.
Maximum Speed: 100 mph.
Doors: Power operated sliding plug.
Couplers: Tightlock.
Bogies: ACR.
Multiple Working: Within Class.

67601–67630. DMC(A). Dia. EA302. Alstom Birmingham 1998–2000. 12/63. 45.2 t.
67701–67730. DMC(B). Dia. EA303. Alstom Birmingham 1998–2000. 12/63. 45.2 t.
74001–74030. PTS. Dia. EH250. Alstom Birmingham 1998–2000. –/49 1TD 2W. 33.3 t.
74101–74130. MS. Dia. EC226. Alstom Birmingham 1998–2000. –/75 1T. 40.6 t.

8001	**U**	P	*SW*	WD	67601	74001	74101	67701
8002	**SW**	P	*SW*	WD	67602	74002	74102	67702
8003	**SW**	P	*SW*	WD	67603	74003	74103	67703
8004	**SW**	P	*SW*	WD	67604	74004	74104	67704
8005		P	*SW*		67605	74005	74105	67705
8006		P	*SW*		67606	74006	74106	67706
8007		P	*SW*		67607	74007	74107	67707
8008		P	*SW*		67608	74008	74108	67708
8009		P	*SW*		67609	74009	74109	67709
8010		P	*SW*		67610	74010	74110	67710
8011		P	*SW*		67611	74011	74111	67711
8012		P	*SW*		67612	74012	74112	67712
8013		P	*SW*		67613	74013	74113	67713
8014		P	*SW*		67614	74014	74114	67714
8015		P	*SW*		67615	74015	74115	67715
8016		P	*SW*		67616	74016	74116	67716
8017		P	*SW*		67617	74017	74117	67717
8018		P	*SW*		67618	74018	74118	67718
8019		P	*SW*		67619	74019	74119	67719
8020		P	*SW*		67620	74020	74120	67720
8021		P	*SW*		67621	74021	74121	67721
8022		P	*SW*		67622	74022	74122	67722
8023		P	*SW*		67623	74023	74123	67723
8024		P	*SW*		67624	74024	74124	67724
8025		P	*SW*		67625	74025	74125	67724
8026		P	*SW*		67626	74026	74126	67726
8027		P	*SW*		67627	74027	74127	67727
8028		P	*SW*		67628	74028	74128	67728
8029		P	*SW*		67629	74029	74129	67729
8030		P	*SW*		67630	74030	74130	67730

CLASS 488 2- or 3-Car Express Trailer Unit

Various formations, see below. Gangwayed throughout. Air conditioned.
Dimensions: 20.38 x 2.84 x 3.79 m.
Maximum Speed: 90 mph.
Doors: Manually operated slam.
Couplings: Buckeye.
Bogies: B4.
Multiple Working: SR type.
Advertising Livery:
- Continental Airlines.

72500–72509. TFH. Dia. EP101. Built as loco-hauled vehicles to Lot No. 30859 Derby 1973–74. Converted BREL Eastleigh 1983–84. 41/– 1T. 35.0 t.
72602–14/6–8/20–44/6/7. TSH. Dia. EP201. Built as loco-hauled vehicles to Lot No. 30860 Derby 1973–74. Converted BREL Eastleigh 1983–84. –/48 1T. 35.0 t.
72615/19/45. TSH. Dia. EP201. Built as loco-hauled vehicles to Lot No. 30846 Derby 1973. Converted BREL Eastleigh 1983–84. –/48 1T. 35.0 t.
72701–72718. TS. Dia. EH290. Built as loco-hauled vehicles to Lot No. 30860 Derby 1973–74. Converted BREL Eastleigh 1983–84. –/48 1T. 35.0 t.

CLASS 488/2. 2-car units. TFH–TSH.

8201	**GX**	P	*GX*	SL	72500	72638
8202	**AL**	P	*GX*	SL	72501	72617
8203	**AL**	P	*GX*	SL	72502	72640
8204	**AL**	P	*GX*	SL	72503	72641
8205	**GX**	P	*GX*	SL	72504	72628
8206	**GX**	P	*GX*	SL	72505	72629
8207	**AL**	P	*GX*	SL	72506	72642
8208	**AL**	P	*GX*	SL	72507	72643
8209	**GX**	P	*GX*	SL	72508	72644
8210	**GX**	P	*GX*	SL	72509	72635

CLASS 488/3. 3-car units. TFH–TS–TSH.

8302	**GX**	P	*GX*	SL	72602	72701	72604
8303	**GX**	P	*GX*	SL	72603	72702	72608
8304	**AL**	P	*GX*	SL	72606	72703	72611
8305	**AL**	P	*GX*	SL	72605	72704	72609
8306	**GX**	P	*GX*	SL	72607	72705	72610
8307	**GX**	P	*GX*	SL	72612	72706	72613
8308	**GX**	P	*GX*	SL	72614	72707	72615
8309	**GX**	P	*GX*	SL	72616	72708	72639
8310	**AL**	P	*GX*	SL	72618	72709	72619
8311	**GX**	P	*GX*	SL	72620	72710	72621
8312	**GX**	P	*GX*	SL	72622	72711	72623
8313	**GX**	P	*GX*	SL	72624	72712	72625
8314	**AL**	P	*GX*	SL	72626	72713	72627
8315	**GX**	P	*GX*	SL	72636	72714	72645
8316	**GX**	P	*GX*	SL	72630	72715	72631
8317	**GX**	P	*GX*	SL	72632	72716	72633
8318	**GX**	P	*GX*	SL	72634	72717	72637
8319	**AL**	P	*GX*	SL	72646	72718	72647

CLASS 489 Gatwick Luggage Van

DMLV. Gangwayed at non-driving end only. Operate with Class 488.
Traction Motors: Two English Electric 507 of 185 kW each.
Dimensions: 20.45 x 2.82 x 3.86 m. **Doors**: Manually operated slam.
Maximum Speed: 90 mph. **Bogies**: B4.
Couplings: Buckeye. **Multiple Working**: SR type.

68500–68509. DMLV. Dia. EB501. Built as DMBS to Lot No. 30452 Eastleigh 1959. Converted BREL Eastleigh 1983–84. 40.5 t.

9101	**GX**	P	*GX*	SL	68500
9102	**GX**	P	*GX*	SL	68501
9103	**GX**	P	*GX*	SL	68502
9104	**GX**	P	*GX*	SL	68503
9105	**GX**	P	*GX*	SL	68504
9106	**GX**	P	*GX*	SL	68505
9107	**GX**	P	*GX*	SL	68506
9108	**GX**	P	*GX*	SL	68507
9109	**GX**	P	*GX*	SL	68508
9110	**GX**	P	*GX*	SL	68509

CLASS 424 Adtranz 'Classic' Prototype

DTS. Gangwayed within unit.
Dimensions: **Doors**: Power operated sliding.
Maximum Speed: 90 mph. **Bogies**: B5 (SR).
Couplers: Tightlock. **Multiple Working**:
Non Standard Livery:
- Silver with black window surrounds.

76112. DTS. Dia. EE280. Built as DTCso to Lot No. 30741 York 1963–66. Rebuilt Adtranz Derby 1997. –/77. 34.0 t.

424 001	**0**	A		ZD(S)	76112

CLASS 456 2-Car Unit

DMS–DTS. Gangwayed within unit. Disc brakes.
Traction Motors: Two GEC507-20J of 185 kW each.
Dimensions: 19.95 x 2.82 x . m.
Maximum Speed: 75 mph. **Doors**: Power operated sliding.
Couplers: Tightlock. **Bogies**: BREL P7/T3.
Multiple Working: Classes 455–457, 507 & 508.

64735–64758. DMS. Dia. EA267. Lot No. 31073 BREL York 1990–91. –/79. 41.1 t.
78250–78273. DTS. Dia. EE276. Lot No. 31074 BREL York 1990–91. –/73. 31.4 t.

456 001	**N**	P	*SC*	SU	64735	78250
456 002	**N**	P	*SC*	SU	64736	78251
456 003	**N**	P	*SC*	SU	64737	78252
456 004	**N**	P	*SC*	SU	64738	78253

456 005	**N**	P	*SC*	SU	64739	78254
456 006	**N**	P	*SC*	SU	64740	78255
456 007	**N**	P	*SC*	SU	64741	78256
456 008	**N**	P	*SC*	SU	64742	78257
456 009	**N**	P	*SC*	SU	64743	78258
456 010	**N**	P	*SC*	SU	64744	78259
456 011	**N**	P	*SC*	SU	64745	78260
456 012	**N**	P	*SC*	SU	64746	78261
456 013	**N**	P	*SC*	SU	64747	78262
456 014	**N**	P	*SC*	SU	64748	78263
456 015	**N**	P	*SC*	SU	64749	78264
456 016	**N**	P	*SC*	SU	64750	78265
456 017	**N**	P	*SC*	SU	64751	78266
456 018	**N**	P	*SC*	SU	64752	78267
456 019	**N**	P	*SC*	SU	64753	78268
456 020	**N**	P	*SC*	SU	64754	78269
456 021	**N**	P	*SC*	SU	64755	78270
456 022	**N**	P	*SC*	SU	64756	78271
456 023	**N**	P	*SC*	SU	64757	78272
456 024	**CX**	P	*SC*	SU	64758	78273

Name (carried on DTS):

456 024 Sir Cosmo Bonsor

CLASS 460 Alstom 'Juniper' 8-Gat Express Unit

DMF–TF–TC–MS(A)–MS(B)–TS–MS(C)–DMS. Gangwayed within unit. Disc and regenerative brakes.
Traction Motors: Two Alstom Onix 800 of 270 kW each per power car.
Dimensions: 21.01 (DMF & DMS) or 19.94 (other cars) x 2.80 x . m.
Maximum Speed: 100 mph.
Doors: Power operated sliding plug.
Couplers: Scharfenberg.
Bogies: ACR.
Multiple Working:

67901–67908. DMF. Dia. EA101. Alstom Birmingham. 1999–2000. 10/– . 42.6 t.
67911–67918. DMS. Dia. EA274. Alstom Birmingham 1999–2000. –/56. 45.3 t.
74401–74408. TF. Dia. EH161. Alstom Birmingham 1999–2000. 28/– 1TD 1W. 33.5 t.
74411–74418. TC. Dia. EH364. Alstom Birmingham 1999–2000. 9/42 1T. 34.9 t.
74421–74428. MS(A). Dia. EC227. Alstom Birmingham 1999–2000. –/60. 42.5 t.
74431–74438. MS(B). Dia. EC228 . Alstom Birmingham 1999–2000. –/60. 42.5 t.
74441–74448. TS. Dia. EH251. Alstom Birmingham 1999–2000. –/38 1TD 1W. 35.2 t.
74451–74458. MS(C). Dia. EC229 . Alstom Birmingham 1999–2000. –/60. 40.5 t.

460 001	**U**	P	*GX*	SL	67901	74401	74411	74421	74431	74441	74451	67911
460 002	**U**	P	*GX*	SL	67902	74402	74412	74422	74432	74442	74452	67912
460 003	**GX**	P	*GX*	SL	67903	74403	74413	74423	74433	74443	74453	67913
460 004	**GX**	P	*GX*	SL	67904	74404	74414	74424	74434	74444	74454	67914
460 005		P	*GX*		67905	74405	74415	74425	74435	74445	74455	67915
460 006		P	*GX*		67906	74406	74416	74426	74436	74446	74456	67916
460 007		P	*GX*		67907	74407	74417	74427	74437	74447	74457	67917
460 008		P	*GX*		67908	74408	74418	74428	74438	74448	74458	67918

CLASS 465 'Networker' 4-car Unit

DMS–TS(A)–TS(B)–DMS. Gangwayed within unit. Disc, rheostatic and regenerative brakes.
Traction Motors: Four Brush TIM 970 or GEC-Alsthom G352BY of 280 kW each per motor car.
Dimensions: 20.89 (DMS) or 20.06 (other cars) x 2.81 x 3.77 m.
Maximum Speed: 75 mph. **Doors:** Power operated sliding plug.
Couplers: Tightlock. **Bogies:** BREL P3/T3.
Multiple Working: Classes 365, 465 & 466. Couplers within units on Class 465/2 cars are not compatible with Classes 465/0 and 465/1.

64759–64858. DMS. Dia. EA268. Lot No. 31100 BREL York 1992–93. –/86. 39.2 t.
65700–65799. DMS. Dia. EA269. Lot No. 31103 GEC-Alsthom Birmingham 1992–93. –/86. 38.9 t.
65800–65893. DMS. Dia. EA268. Lot No. 31130 ABB York 1993–94. –/86. 39.0 t.
72028–72126 (Even numbers). TS(A). Dia. EH293. Lot No. 31102 BREL York 1992–93. –/90. 30.4 t.
72029–72127 (Odd numbers). TS(B). Dia. EH292. Lot No. 31101 BREL York 1992–93. –/86. 30.5 t.
72719–72817 (Odd numbers). TS(A). Dia. EH294. Lot No. 31104 GEC-Alsthom Birmingham 1992–93. –/86. 30.2 t.
72720–72818 (Even numbers). TS(B). Dia. EH295. Lot No. 31105 GEC-Alsthom Birmingham 1992–93. –/90. 29.1 t.
72900–72992 (Even numbers). TS(A). Dia. EH293. Lot No. 31132 ABB York 1993–94. –/90. 29.5 t.
72901–72993 (Odd numbers). TS(B). Dia. EH294. Lot No. 31131 ABB York 1993–94. –/86. 30.2 t.

Class 465/0. Built by ABB. Brush traction motors.

465 001	**CS**	H	*SE*	SG	64759	72028	72029	64809
465 002	**CS**	H	*SE*	SG	64760	72030	72031	64810
465 003	**CS**	H	*SE*	SG	64761	72032	72033	64811
465 004	**NT**	H	*SE*	SG	64762	72034	72035	64812
465 005	**NT**	H	*SE*	SG	64763	72036	72913	64813
465 006	**CS**	H	*SE*	SG	64764	72038	72039	64814
465 007	**CS**	H	*SE*	SG	64765	72040	72041	64815
465 008	**CS**	H	*SE*	SG	64766	72042	72043	64816
465 009	**CS**	H	*SE*	SG	64767	72044	72045	64817
465 010	**CS**	H	*SE*	SG	64768	72046	72047	64818
465 011	**CS**	H	*SE*	SG	64769	72048	72049	64819
465 012	**CS**	H	*SE*	SG	64770	72050	72051	64820
465 013	**CS**	H	*SE*	SG	64771	72052	72053	64821
465 014	**CS**	H	*SE*	SG	64772	72054	72055	64822
465 015	**CS**	H	*SE*	SG	64773	72056	72057	64823
465 016	**CS**	H	*SE*	SG	64774	72058	72059	64824
465 017	**CS**	H	*SE*	SG	64775	72060	72061	64825
465 018	**CS**	H	*SE*	SG	64776	72062	72063	64826
465 019	**CS**	H	*SE*	SG	64777	72064	72065	64827
465 020	**CS**	H	*SE*	SG	64778	72066	72067	64828

465 021	**NT**	H	*SE*	SG	64779	72068	72069	64829
465 022	**NT**	H	*SE*	SG	64780	72070	72071	64830
465 023	**NT**	H	*SE*	SG	64781	72072	72073	64831
465 024	**NT**	H	*SE*	SG	64782	72074	72075	64832
465 025	**NT**	H	*SE*	SG	64783	72076	72077	64833
465 026	**NT**	H	*SE*	SG	64784	72078	72079	64834
465 027	**NT**	H	*SE*	SG	64785	72080	72081	64835
465 028	**NT**	H	*SE*	SG	64786	72082	72083	64836
465 029	**NT**	H	*SE*	SG	64787	72084	72085	64837
465 030	**NT**	H	*SE*	SG	64788	72086	72087	64838
465 031	**NT**	H	*SE*	SG	64789	72088	72089	64839
465 032	**NT**	H	*SE*	SG	64790	72090	72091	64840
465 033	**NT**	H	*SE*	SG	64791	72092	72093	64841
465 034	**NT**	H	*SE*	SG	64792	72094	72095	64842
465 035	**NT**	H	*SE*	SG	64793	72096	72097	64843
465 036	**NT**	H	*SE*	SG	64794	72098	72099	64844
465 037	**NT**	H	*SE*	SG	64795	72100	72101	64845
465 038	**NT**	H	*SE*	SG	64796	72102	72103	64846
465 039	**NT**	H	*SE*	SG	64797	72104	72105	64847
465 040	**NT**	H	*SE*	SG	64798	72106	72107	64848
465 041	**NT**	H	*SE*	SG	64799	72108	72109	64849
465 042	**NT**	H	*SE*	SG	64800	72110	72111	64850
465 043	**NT**	H	*SE*	SG	64801	72112	72113	64851
465 044	**NT**	H	*SE*	SG	64802	72114	72115	64852
465 045	**NT**	H	*SE*	SG	64803	72116	72117	64853
465 046	**NT**	H	*SE*	SG	64804	72118	72119	64854
465 047	**NT**	H	*SE*	SG	64805	72120	72121	64855
465 048	**NT**	H	*SE*	SG	64806	72122	72123	64856
465 049	**NT**	H	*SE*	SG	64807	72124	72125	64857
465 050	**NT**	H	*SE*	SG	64808	72126	72127	64858

Class 465/1. Built by ABB. Brush traction motors.

465 151	**NT**	H	*SE*	SG	65800	72900	72901	65847
465 152	**NT**	H	*SE*	SG	65801	72902	72903	65848
465 153	**NT**	H	*SE*	SG	65802	72904	72905	65849
465 154	**NT**	H	*SE*	SG	65803	72906	72907	65850
465 155	**NT**	H	*SE*	SG	65804	72908	72909	65851
465 156	**NT**	H	*SE*	SG	65805	72910	72911	65852
465 157	**NT**	H	*SE*	SG	65816	72912	72037	65853
465 158	**NT**	H	*SE*	SG	65807	72914	72915	65854
465 159	**NT**	H	*SE*	SG	65808	72916	72917	65855
465 160	**NT**	H	*SE*	SG	65809	72918	72919	65856
465 161	**NT**	H	*SE*	SG	65810	72920	72921	65857
465 162	**NT**	H	*SE*	SG	65811	72922	72923	65858
465 163	**NT**	H	*SE*	SG	65812	72924	72925	65859
465 164	**NT**	H	*SE*	SG	65813	72926	72927	65860
465 165	**NT**	H	*SE*	SG	65814	72928	72929	65861
465 166	**NT**	H	*SE*	SG	65815	72930	72931	65862
465 167	**NT**	H	*SE*	SG	65806	72932	72933	65863
465 168	**NT**	H	*SE*	SG	65817	72934	72935	65864
465 169	**NT**	H	*SE*	SG	65818	72936	72937	65865

465 170	**NT**	H	*SE*	SG	65819	72938	72939	65866
465 171	**NT**	H	*SE*	SG	65820	72940	72941	65867
465 172	**NT**	H	*SE*	SG	65821	72942	72943	65868
465 173	**NT**	H	*SE*	SG	65822	72944	72945	65869
465 174	**NT**	H	*SE*	SG	65823	72946	72947	65870
465 175	**NT**	H	*SE*	SG	65824	72948	72949	65871
465 176	**NT**	H	*SE*	SG	65825	72950	72951	65872
465 177	**NT**	H	*SE*	SG	65826	72952	72953	65873
465 178	**NT**	H	*SE*	SG	65827	72954	72955	65874
465 179	**NT**	H	*SE*	SG	65828	72956	72957	65875
465 180	**NT**	H	*SE*	SG	65829	72958	72959	65876
465 181	**NT**	H	*SE*	SG	65830	72960	72961	65877
465 182	**NT**	H	*SE*	SG	65831	72962	72963	65878
465 183	**NT**	H	*SE*	SG	65832	72964	72965	65879
465 184	**NT**	H	*SE*	SG	65833	72966	72967	65880
465 185	**NT**	H	*SE*	SG	65834	72968	72969	65881
465 186	**NT**	H	*SE*	SG	65835	72970	72971	65882
465 187	**NT**	H	*SE*	SG	65836	72972	72973	65883
465 188	**NT**	H	*SE*	SG	65837	72974	72975	65884
465 189	**NT**	H	*SE*	SG	65838	72976	72977	65885
465 190	**NT**	H	*SE*	SG	65839	72978	72979	65886
465 191	**NT**	H	*SE*	SG	65840	72980	72981	65887
465 192	**NT**	H	*SE*	SG	65841	72982	72983	65888
465 193	**NT**	H	*SE*	SG	65842	72984	72985	65889
465 194	**NT**	H	*SE*	SG	65843	72986	72987	65890
465 195	**NT**	H	*SE*	SG	65844	72988	72989	65891
465 196	**NT**	H	*SE*	SG	65845	72990	72991	65892
465 197	**NT**	H	*SE*	SG	65846	72992	72993	65893

Class 465/2. Built by GEC-Alsthom. GEC-Alsthom traction motors.

465 201	**NT**	A	*SE*	SG	65700	72719	72720	65750
465 202	**NT**	A	*SE*	SG	65701	72721	72722	65751
465 203	**NT**	A	*SE*	SG	65702	72723	72724	65752
465 204	**NT**	A	*SE*	SG	65703	72725	72726	65753
465 205	**NT**	A	*SE*	SG	65704	72727	72728	65754
465 206	**NT**	A	*SE*	SG	65705	72729	72730	65755
465 207	**NT**	A	*SE*	SG	65706	72731	72732	65756
465 208	**NT**	A	*SE*	SG	65707	72733	72734	65757
465 209	**NT**	A	*SE*	SG	65708	72735	72736	65758
465 210	**NT**	A	*SE*	SG	65709	72737	72738	65759
465 211	**NT**	A	*SE*	SG	65710	72739	72740	65760
465 212	**NT**	A	*SE*	SG	65711	72741	72742	65761
465 213	**NT**	A	*SE*	SG	65712	72743	72744	65762
465 214	**NT**	A	*SE*	SG	65713	72745	72746	65763
465 215	**NT**	A	*SE*	SG	65714	72747	72748	65764
465 216	**NT**	A	*SE*	SG	65715	72749	72750	65765
465 217	**NT**	A	*SE*	SG	65716	72751	72752	65766
465 218	**NT**	A	*SE*	SG	65717	72753	72754	65767
465 219	**NT**	A	*SE*	SG	65718	72755	72756	65768
465 220	**NT**	A	*SE*	SG	65719	72757	72758	65769
465 221	**NT**	A	*SE*	SG	65720	72759	72760	65770

465 222	**NT**	A	*SE*	SG	65721	72761	72762	65771
465 223	**NT**	A	*SE*	SG	65722	72763	72764	65772
465 224	**NT**	A	*SE*	SG	65723	72765	72766	65773
465 225	**NT**	A	*SE*	SG	65724	72767	72768	65774
465 226	**NT**	A	*SE*	SG	65725	72769	72770	65775
465 227	**NT**	A	*SE*	SG	65726	72771	72772	65776
465 228	**NT**	A	*SE*	SG	65727	72773	72774	65777
465 229	**NT**	A	*SE*	SG	65728	72775	72776	65778
465 230	**NT**	A	*SE*	SG	65729	72777	72778	65779
465 231	**NT**	A	*SE*	SG	65730	72779	72780	65780
465 232	**NT**	A	*SE*	SG	65731	72781	72782	65781
465 233	**NT**	A	*SE*	SG	65732	72783	72784	65782
465 234	**NT**	A	*SE*	SG	65733	72785	72786	65783
465 235	**NT**	A	*SE*	SG	65734	72787	72788	65784
465 236	**NT**	A	*SE*	SG	65735	72789	72790	65785
465 237	**NT**	A	*SE*	SG	65736	72791	72792	65786
465 238	**NT**	A	*SE*	SG	65737	72793	72794	65787
465 239	**NT**	A	*SE*	SG	65738	72795	72796	65788
465 240	**NT**	A	*SE*	SG	65739	72797	72798	65789
465 241	**NT**	A	*SE*	SG	65740	72799	72800	65790
465 242	**NT**	A	*SE*	SG	65741	72801	72802	65791
465 243	**NT**	A	*SE*	SG	65742	72803	72804	65792
465 244	**NT**	A	*SE*	SG	65743	72805	72806	65793
465 245	**NT**	A	*SE*	SG	65744	72807	72808	65794
465 246	**NT**	A	*SE*	SG	65745	72809	72810	65795
465 247	**NT**	A	*SE*	SG	65746	72811	72812	65796
465 248	**NT**	A	*SE*	SG	65747	72813	72814	65797
465 249	**NT**	A	*SE*	SG	65748	72815	72816	65798
465 250	**NT**	A	*SE*	SG	65749	72817	72818	65799

CLASS 466 Networker 2-Car Unit

DMS–DTS. Gangwayed within unit. Disc, rheostatic and regenerative brakes. 466 017 has experimental 2 + 2 seating.
Traction Motors: Four GEC-Alsthom G352BY of 280 kW each.
Dimensions: 20.89 (DMS) or 20.06 (DTS) x 2.81 x 3.77 m.
Maximum Speed: 75 mph. **Doors:** Power operated sliding plug.
Couplers: Tightlock. **Bogies:** BREL P3/T3.
Multiple Working: Classes 365, 465 & 466.

64860–64902. DMS. Dia. EA271. Lot No. 31128 GEC-Alsthom Birmingham 1993–94. –/86 (* –/72). 38.8 t.
78312–78354. DTS. Dia. EE279. Lot No. 31129 GEC-Alsthom Birmingham 1993–94. –/82 (* –/68). 33.2 t.

466 001	**NT**	A	*SE*	SG	64860	78312
466 002	**NT**	A	*SE*	SG	64861	78313
466 003	**NT**	A	*SE*	SG	64862	78314
466 004	**NT**	A	*SE*	SG	64863	78315
466 005	**NT**	A	*SE*	SG	64864	78316
466 006	**NT**	A	*SE*	SG	64865	78317

466 007		**NT**	A	*SE*	SG	64866	78318
466 008		**NT**	A	*SE*	SG	64867	78319
466 009		**NT**	A	*SE*	SG	64868	78320
466 010		**NT**	A	*SE*	SG	64869	78321
466 011		**NT**	A	*SE*	SG	64870	78322
466 012		**NT**	A	*SE*	SG	64871	78323
466 013		**NT**	A	*SE*	SG	64872	78324
466 014		**NT**	A	*SE*	SG	64873	78325
466 015		**NT**	A	*SE*	SG	64874	78326
466 016		**NT**	A	*SE*	SG	64875	78327
466 017	*	**NT**	A	*SE*	SG	64876	78328
466 018		**NT**	A	*SE*	SG	64877	78329
466 019		**NT**	A	*SE*	SG	64878	78330
466 020		**NT**	A	*SE*	SG	64879	78331
466 021		**NT**	A	*SE*	SG	64880	78332
466 022		**NT**	A	*SE*	SG	64881	78333
466 023		**NT**	A	*SE*	SG	64882	78334
466 024		**NT**	A	*SE*	SG	64883	78335
466 025		**NT**	A	*SE*	SG	64884	78336
466 026		**NT**	A	*SE*	SG	64885	78337
466 027		**NT**	A	*SE*	SG	64886	78338
466 028		**NT**	A	*SE*	SG	64887	78339
466 029		**NT**	A	*SE*	SG	64888	78340
466 030		**NT**	A	*SE*	SG	64889	78341
466 031		**NT**	A	*SE*	SG	64890	78342
466 032		**NT**	A	*SE*	SG	64891	78343
466 033		**NT**	A	*SE*	SG	64892	78344
466 034		**NT**	A	*SE*	SG	64893	78345
466 035		**NT**	A	*SE*	SG	64894	78346
466 036		**NT**	A	*SE*	SG	64895	78347
466 037		**NT**	A	*SE*	SG	64896	78348
466 038		**NT**	A	*SE*	SG	64897	78349
466 039		**NT**	A	*SE*	SG	64898	78350
466 040		**NT**	A	*SE*	SG	64899	78351
466 041		**NT**	A	*SE*	SG	64900	78352
466 042		**NT**	A	*SE*	SG	64901	78353
466 043		**NT**	A	*SE*	SG	64902	78354

CLASS 507 Merseyrail 3-Car Unit

BDMS–TS–DMS. Gangwayed within unit. End doors. Disc & rheostatic brakes.
Traction Motors: Four GEC G310AZ of 82.125 kW per power car.
Dimensions: 20.02 (BDMS & DMS) or 19.92 (TS) x 2.82 x 3.58 m.
Maximum Speed: 75 mph. **Doors**: Power operated sliding.
Couplers: Tightlock. **Bogies**: BREL BX1.
Multiple Working: Classes 507 & 508.
Note: Fitted with de-icing equipment.

64367–64399. BDMS. Dia. EI202. Lot No. 30906 York 1978–80. –/74. 37.1 t.
64405–64437. DMS. Dia. EA201. Lot No. 30908 York 1978–80. –/74. 35.6 t.
71342–71374. TS. Dia. EH205. Lot No. 30907 York 1978–80. –/82. 25.6 t.

507 001	**MT**	A	*ME*	BD	64367	71342	64405
507 002	**MT**	A	*ME*	BD	64368	71343	64406
507 003	**MT**	A	*ME*	BD	64369	71344	64407
507 004	**MT**	A	*ME*	BD	64388	71345	64408
507 005	**MT**	A	*ME*	BD	64371	71346	64409
507 006	**MT**	A	*ME*	BD	64372	71347	64410
507 007	**MT**	A	*ME*	BD	64373	71348	64411
507 008	**MT**	A	*ME*	BD	64374	71349	64412
507 009	**MT**	A	*ME*	BD	64375	71350	64413
507 010	**MT**	A	*ME*	BD	64376	71351	64414
507 011	**MT**	A	*ME*	BD	64377	71352	64415
507 012	**MT**	A	*ME*	BD	64378	71353	64416
507 013	**MT**	A	*ME*	BD	64379	71354	64417
507 014	**MT**	A	*ME*	BD	64380	71355	64418
507 015	**MT**	A	*ME*	BD	64381	71356	64419
507 016	**MT**	A	*ME*	BD	64382	71357	64420
507 017	**MT**	A	*ME*	BD	64383	71358	64421
507 018	**MT**	A	*ME*	BD	64384	71359	64422
507 019	**MT**	A	*ME*	BD	64385	71360	64423
507 020	**MT**	A	*ME*	BD	64386	71361	64424
507 021	**MT**	A	*ME*	BD	64387	71362	64425
507 023	**MT**	A	*ME*	BD	64389	71364	64427
507 024	**MT**	A	*ME*	BD	64390	71365	64428
507 025	**MT**	A	*ME*	BD	64391	71366	64429
507 026	**MT**	A	*ME*	BD	64392	71367	64430
507 027	**MT**	A	*ME*	BD	64393	71368	64431
507 028	**MT**	A	*ME*	BD	64394	71369	64432
507 029	**MT**	A	*ME*	BD	64395	71370	64433
507 030	**MT**	A	*ME*	BD	64396	71371	64434
507 031	**MT**	A	*ME*	BD	64397	71372	64435
507 032	**MT**	A	*ME*	BD	64398	71373	64436
507 033	**MT**	A	*ME*	BD	64399	71374	64437

CLASS 508 3-Car Unit

DMS–TS–BDMS. Gangwayed within unit. End doors. Disc & rheostatic brakes.
Traction Motors: Four GEC G310AZ of 82.125 kW per power car.
Dimensions: 20.02 (DMS & BDMS) or 19.92 (TS) x 2.82 x 3.58 m.
Maximum Speed: 75 mph. **Doors:** Power operated sliding.
Couplers: Tightlock. **Bogies:** BREL BX1.
Multiple Working: Classes 507 & 508.

64649–64691. DMS. Dia. EA208 (*EA211). Lot No. 30979 BREL York 1979–80 (* Facelifted 1998–99 by Wessex Traincare/Alstom Eastleigh). –/68 (*–/66). 36.2 t.

64692–64734. BDMS. Dia. EI203 (*EI204). Lot No. 30981 BREL York 1979–80 (Facelifted 1998–99 by Wessex Traincare/Alstom Eastleigh). –/68 (*–/74). 36.6 t.

71483–71525. TS. Dia. EH218 (*EH246). Lot No. 30980 BREL York 1979–80 (Facelifted 1998–99 by Wessex Traincare/Alstom Eastleigh). –/86 (*–/79). 26.7 t.

Class 508/1. Standard design.

508 102	**MT**	A		WK(S)	64650	71484	64693
508 103	**MT**	A	*ME*	BD	64651	71485	64694
508 104	**MT**	A	*ME*	BD	64652	71486	64695
508 108	**MT**	A		NB(S)	64656	71490	64699
508 110	**MT**	A		KK(S)	64658	71492	64701
508 111	**MT**	A	*ME*	BD	64659	71493	64702
508 112	**MT**	A	*ME*	BD	64660	71494	64703
508 114	**MT**	A	*ME*	BD	64662	71496	64705
508 115	**MT**	A	*ME*	BD	64663	71497	64706
508 117	**MT**	A	*ME*	BD	64665	71499	64708
508 118	**MT**	A		BD(S)	64666	71500	64709
508 120	**MT**	A		KK(S)	64668	71502	64711
508 122	**MT**	A		KK(S)	64670	71504	64713
508 123	**MT**	A		NB(S)	64671	71505	64714
508 124	**MT**	A	*ME*	BD	64672	71506	64715
508 125	**MT**	A	*ME*	BD	64673	71507	64716
508 126	**MT**	A	*ME*	BD	64674	71508	64717
508 127	**MT**	A	*ME*	BD	64675	71509	64718
508 128	**MT**	A	*ME*	BD	64676	71510	64719
508 130	**MT**	A	*ME*	BD	64678	71512	64721
508 131	**MT**	A		KK(S)	64679	71513	64722
508 134	**MT**	A	*ME*	BD	64682	71516	64725
508 135	**MT**	A		NB(S)	64683	71517	64726
508 136	**MT**	A	*ME*	BD	64684	71518	64727
508 137	**MT**	A	*ME*	BD	64685	71519	64728
508 138	**MT**	A	*ME*	BD	64686	71520	64729
508 139	**MT**	A	*ME*	BD	64687	71521	64730
508 140	**MT**	A	*ME*	BD	64688	71522	64731
508 141	**MT**	A	*ME*	BD	64689	71523	64732
508 142	**MT**	A		WK(S)	64690	71524	64733
508 143	**MT**	A	*ME*	BD	64691	71525	64734

Class 508/2. Facelifted units for Connex South Eastern.

508 201	*	**CX**	A	*SE*	GI	64649	71483	64692
508 202	*	**CX**	A	*SE*	GI	64653	71487	64696
508 203	*	**CX**	A	*SE*	GI	64654	71488	64697
508 204	*	**CX**	A	*SE*	GI	64655	71489	64698
508 205	*	**CX**	A	*SE*	GI	64657	71491	64700
508 206	*	**CX**	A	*SE*	GI	64661	71495	64714
508 207	*	**CX**	A	*SE*	GI	64664	71498	64707
508 208	*	**CX**	A	*SE*	GI	64667	71501	64710
508 209	*	**CX**	A	*SE*	GI	64669	71503	64712
508 210	*	**CX**	A	*SE*	GI	64677	71511	64720
508 211	*	**CX**	A	*SE*	GI	64680	71514	64723
508 212	*	**CX**	A	*SE*	GI	64681	71515	64724

3. EUROSTAR UNITS (CLASS 373)

Eurostar units are used on services between Britain and Continental Europe via the Channel Tunnel.

Each train consists of two Eurostar units coupled, with a power car at each driving end. Services starting from/terminating at London Waterloo International are formed of two 9-car unitss coupled, whilst those to/from other British destinations (yet to commence) will be formed of two 7-car units coupled. All units are articulated with an extra motor bogie on the coach adjacent to the power car.

DM–MS–4TS–RB–2TF–TBF or DM–MS–3TS–RB–TF–TBF. Gangwayed within pair of units. Air conditioned.
Supply Systems: 25 kV a.c. 50 Hz overhead or 3000 V d.c. overhead or 750 V d.c. third rail (* also equipped for 1500 V d.c. overhead operation).
Wheel Arrangement: Bo–Bo + Bo–2–2–2–2–2–2–2–2–2.
Length: 22.15 m (DM), 21.85 m (MSOL & TBFOL), 18.70 m (other cars).
Maximum Speed: 300 km/h.
Built: 1992-93 by GEC-Alsthom/Brush/ANF/De Dietrich/BN Construction/ACEC.
Note: DM vehicles carry the set numbers indicated below.

10-Car Sets. Built for services starting from/terminating at London Waterloo. Individual vehicles in each set are allocated numbers 7xxxx0 + 7xxxx1 + 7xxxx2 + 7xxxx3 + 7xxxx4 + 7xxxx5 + 7xxxx6 + 7xxxx8 + 7xxxx9, where xxxx denotes the set number.

Advertising Liveries:
- 3305/06 'The Beatles Yellow Submarine'.

73xxx0 series. DM. Dia. LA501. Lot No. 31118 1992–95. 68.5 t.
73xxx1 series. MS. Dia. LB202. Lot No. 31119 1992–95. –/48 2T. 44.6 t.
73xxx2 series. TS. Dia. LC202. Lot No. 31120 1992–95. –/58 1T. 28.1 t.
73xxx3 series. TS. Dia. LD202. Lot No. 31121 1992–95. –/58 2T. 29.7 t.
73xxx4 series. TS. Dia. LE202. Lot No. 31122 1992–95. –/58 1T. 28.3 t.
73xxx5 series. TS. Dia. LF202. Lot No. 31123 1992–95. –/58 2T. 29.2 t.
73xxx6 series. RB. Dia. LG502. Lot No.31124 1992–95. 31.1 t.
73xxx7 series. TF. Dia. LH102. Lot No. 31125 1992–95. 39/– 1T. 29.6 t.
73xxx8 series. TF. Dia. LJ102. Lot No. 31126 1992–95. 39/– 1T. 32.2 t.
73xxx9 series.TBF. Dia. LK102. Lot No. 31127 1992–95. 25/– 1TD. 39.4 t.

3001	**EU**	EU	*EU*	NP
3002	**EU**	EU	*EU*	NP
3003	**EU**	EU	*EU*	NP
3004	**EU**	EU	*EU*	NP
3005	**AL**	EU	*EU*	NP
3006	**AL**	EU	*EU*	NP
3007	**EU**	EU	*EU*	NP
3008	**EU**	EU	*EU*	NP
3009	**EU**	EU	*EU*	NP
3010	**EU**	EU	*EU*	NP
3011	**EU**	EU	*EU*	NP
3012	**EU**	EU	*EU*	NP
3013	**EU**	EU	*EU*	NP
3014	**EU**	EU	*EU*	NP
3015	**EU**	EU	*EU*	NP
3016	**EU**	EU	*EU*	NP
3017	**EU**	EU	*EU*	NP
3018	**EU**	EU	*EU*	NP
3019	**EU**	EU	*EU*	NP
3020	**EU**	EU	*EU*	NP
3021	**EU**	EU	*EU*	NP
3022	**EU**	EU	*EU*	NP
3101	**EU**	SB	*EU*	FF
3102	**EU**	SB	*EU*	FF

3103	**EU**	SB	*EU*	FF	3213	**EU**	SF	*EU*	LY
3104	**EU**	SB	*EU*	FF	3214	**EU**	SF	*EU*	LY
3105	**EU**	SB	*EU*	FF	3215*	**EU**	SF	*EU*	LY
3106	**EU**	SB	*EU*	FF	3216*	**EU**	SF	*EU*	LY
3107	**EU**	SB	*EU*	FF	3217	**EU**	SF	*EU*	LY
3108	**EU**	SB	*EU*	FF	3218	**EU**	SF	*EU*	LY
3201 *	**EU**	SF	*EU*	LY	3219	**EU**	SF	*EU*	LY
3202 *	**EU**	SF	*EU*	LY	3220	**EU**	SF	*EU*	LY
3205	**EU**	SF	*EU*	LY	3221	**EU**	SF	*EU*	LY
3206	**EU**	SF	*EU*	LY	3222	**EU**	SF	*EU*	LY
3207 *	**EU**	SF	*EU*	LY	3223*	**EU**	SF	*EU*	LY
3208 *	**EU**	SF	*EU*	LY	3224*	**EU**	SF	*EU*	LY
3209	**EU**	SF	*EU*	LY	3229*	**EU**	SF	*EU*	LY
3210	**EU**	SF	*EU*	LY	3230*	**EU**	SF	*EU*	LY
3211	**EU**	SF	*EU*	LY	3231	**EU**	SF	*EU*	LY
3212	**EU**	SF	*EU*	LY	3232	**EU**	SF	*EU*	LY

8-Car Sets. Built for Regional Eurostar services. Individual vehicles in each set are allocated numbers 7xxxx0 + 7xxxx1 + 7xxxx3 + 7xxxx2 + 7xxxx5 + 7xxxx6 + 7xxxx7 + 7xxxx9, where xxxx denotes the set number.

733xx0 series. DM. Dia. LA502. 68.5 t.
733xx1 series. MS. Dia. LB203. –/48 1T. 44.6 t.
733xx3 series. TS. Dia. LD203. –/58 2T. 29.7 t.
733xx2 series. TS. Dia. LC203. –/58 1T. 28.1 t.
733xx5 series. TS. Dia. LF203. –/58 1T. 29.2 t.
733xx6 series. RB. Dia. LG503. 31.1 t.
733xx7 series. TF. Dia. LH103. 39/– 1T. 29.6 t.
733xx9 series. TBF. Dia. LK103. 18/– 1TD. 39.4 t.

3301	**EU**	EU	*EU*	NP	3308	**EU**	EU		NP(S)
3302	**EU**	EU	*EU*	NP	3309	**EU**	EU	*EU*	NP
3303	**EU**	EU	*EU*	NP	3310	**EU**	EU	*EU*	NP
3304	**EU**	EU	*EU*	NP	3311	**EU**	EU	*EU*	NP
3305	**EU**	EU	*EU*	NP	3312	**EU**	EU	*EU*	NP
3306	**EU**	EU	*EU*	NP	3313	**EU**	EU		NP(S)
3307	**EU**	EU		NP(S)	3314	**EU**	EU		NP(S)

Spare DM:

3999	**EU**	EU	*EU*	NP

PLATFORM 5 MAIL ORDER

PLATFORM 5 EUROPEAN HANDBOOKS

The Platfom 5 European Handbooks are the most comprehensive guides to the rolling stock of selected European railway administrations available. Each book in the series contains the following information:

- **Locomotives**
- **Preserved Locomotives, Museums and Museum Lines**
- **Railcars and Multiple Units**
- **Technical Data**
- **Depot Allocations (where allocated)**
- **Lists of Depots and Workshops**

Each book is A5 size, thread sewn and includes 32 pages of colour photographs (16 pages in Irish Railways). Benelux Railways and Irish Railways also include details of hauled coaching stock.

The following are currently available:

No.1: Benelux Railways 3rd edition .... £10.50 128 pages. Published 1994
No.3: Austrian Railways 3rd edition .... £10.50 128 pages. Published 1995
No.4: French Railways 3rd edition .... £14.50 176 pages. Published 1999
No.5: Swiss Railways 2nd edition .... £13.50 176 pages. Published 1997
No.6: Italian Railways 1st edition .... £13.50 160 pages. Published 1995
No.7: Irish Railways 1st edition .... £9.95 96 pages. Published 1996

If you would like to be notified when new titles in this series are available, please contact our Mail Order Department. Alternatively, please see our advertisements in Today's Railways magazine for up to date publication information.

To place an order, please follow instructions on page 112.

PLATFORM 5 MAIL ORDER

LIGHT RAIL REVIEW

Light Rail Review is the acclaimed series of Platform 5 books examining developments in light rail around the world. Published since 1989, each volume consists of topical articles from a team of specialist transport writers. Each volume also includes a round up of developments in the UK at time of publication, and a full world list of LRT systems. Much use is made of illustrations and diagrams, the majority of which are reproduced in colour.

Each volume is A4 size, softback with thread sewn binding. The following volumes are still available:

Light Rail Review 3 (Published 1991) £7.50
Manchester, Sheffield, Croydon, Zurich, Amsterdam, Hiroshima

Light Rail Review 4 (Published 1992) £7.50
Manchester, Nantes, Sheffield, Croydon, Romania, USA

Light Rail Review 5 (Published 1993) £7.50
Sheffield, Manchester, Saarbrucken, Leeds, Lille, St. Louis

Light Rail Review 6 (Published 1994) £7.50
Sheffield, Glasgow, Blackpool, Tuen Mun, Strasbourg, Fort Worth

Light Rail Review 7 (Published 1996) £8.95
Midland Metro, Manchester, Sheffield, Japan, Kassel, Ultra LRT

Light Rail Review 8 (Published 1998) £9.50
Midland Metro, Croydon, Amsterdam, Sydney, Sheffield, Portland

To place an order, please follow instructions on page 112.

▲ 357 001 at Adtranz Derby, awaiting delivery to LTS Rail, on 26th August 1999.
Brian Morrison

▼ 365 001 formed the 13.52 Ramsgate–London Victoria service on 3rd February 1998.
Rodney Lissenden

▲ Class 483 units 002 and 004 near Ryde Esplanade with a Shanklin–Ryde Pier Head service on 21st August 1999. **Hugh Ballantyne**

▼ Class 438 trailer unit 417 was hired to Silverlink Train Services for use on the Gospel Oak–Barking line for a short period in the autumn of 1999. It is seen here propelled by 33103 passing Gospel Oak with a crew training run on 19th August 1999. **David Brown**

▲ Class 411/9 unit 1103 forms the 10.07 London Bridge–Tunbridge Wells service at Edenbridge Town on 27th March 1999. **Chris Wilson**

▼ Class 411/5 unit 1534 shunts into the depot sidings at Fratton on 12th July 1999. **Chris Wilson**

▲ Class 421/3 unit 1747 forms the 15.16 London Victoria–Bournemouth, seen passing St. Denys on 23rd July 1999. **Brian Denton**

▼ Class 442 units 2406 and 2412 depart from Basingstoke with the 07.50 Poole–London Waterloo on 26th June 1999. **David Brown**

▲ Class 423 units 3478 and 3501 head a 12-car empty train bound for Brighton on 17th June 1999. **Chris Wilson**

▼ Class 455/7 unit 5734 departs from Clapham Junction with a London Waterloo–Woking service on 22nd October 1997. **Rodney Lissenden**

▲ New Class 460 unit 8004 stands inside Wimbledon T&RSMD on 16th August 1999. **Brian Morrison**

▼ Class 489 luggage van leads an 9-car 'Gatwick Express' formation on the 09.40 Gatwick Airport–London Victoria on 6th August 1999, with Class 73/2 locomotive, 73202 bringing up the rear. **David Brown**

▲ 456 015 departs from London Bridge with the 09.59 London Bridge–Smitham on 11th November 1998. **Rodney Lissenden**

▼ 507 028 leaves Kirkdale with a Liverpool-bound service on 26th June 1999. **Hugh Ballantyne**

▲ Railtrack sandite laying unit 930 102 passes Millbrook on 23rd February 1999.
Brian Denton

▼ Eurostar units 3018 and 3017 pass Paddock Wood with a London Waterloo International–Paris Nord working on 28th August 1999. **Michael J. Collins**

4. SERVICE UNITS

CLASS 316 3-Car Test Unit

BDB–M–DT. Gangwayed within unit. Converted from Class 307. Test bed for Class 323 electrical equipment.
Supply System: 25 kV a.c. 50 Hz overhead or 750 V d.c. third rail.
Traction Motors: Four Holec DMKT52/24 of 146 kW each.
Dimensions: 20.31 (BDB & DT) or 20.18 (M) x 2.82 x 3.86 m.
Maximum Speed: 75 mph.
Doors: Manually operated slam.
Couplings: Buckeye.
Bogies: Gresley ED7/B4.
Multiple Working:

61018. M. Dia. EZ . Lot No. 30203 Eastleigh 1954–56. Converted BR, Derby 1992. . t.
75018. DT. Dia. EZ . Lot No. 30206 Eastleigh 1954–56. Converted BR, Derby 1992. . t.
75118. BDB. Dia. EZ . Lot No. 30205 Eastleigh 1954–56. Converted BR, Derby 1992. . t.

316 997	**BG**	SO	*SO*	ZA	75118	61018	75018

CLASS 930/0 2-Car Sandite & De-icing Units

DMB–DMB. Gangwayed within unit. Converted from Class 405.
Supply System: 750 V d.c. third rail.
Traction Motors: Two English Electric 507 of 185 kW each per car.
Dimensions: 19.05 x 2.74 x 3.99 m.
Maximum Speed: 75 mph.
Doors: Manually operated slam.
Couplings: Buckeye.
Bogies: Central 43 inch.
Multiple Working: SR type.

977586/587/604/605. DMB. Dia. EZ512. Lot No. 3231 SR Eastleigh 1947. 39.0 t.
975588/589/592/595/597–600/602/603. DMB. Dia. EZ512 Lot No. 1060. SR Eastleigh 1941. 39.0 t.
975590/591/596/601. DMB. Dia. EZ512. Lot No. 3384 Eastleigh 1948. 39.0 t.
975593/594. DMB. Dia. EZ512. Lot No. 3618 Eastleigh 1950. 39.0 t.
975896/897. DMB. Dia. EZ512. Lot No. 3506 Eastleigh 1950. 39.0 t.

930 001	**RO**	RK		AF(S)	975596	975605
930 002	**RO**	RK	*RK*	RM	975896	975897
930 003	**RO**	RK	*RK*	SU	975594	975595
930 004	**RO**	RK	*RK*	WD	975586	975587
930 005	**RK**	RK	*RK*	WD	975588	975589
930 006	**RO**	RK	*RK*	WD	975590	975591
930 007	**RO**	RK	*RK*	GI	975592	975593
930 008	**RK**	RK	*RK*	GI	975604	975597
930 009	**N**	RK	*RK*	BI	975598	975599
930 010	**RK**	RK	*RK*	BI	975600	975601
930 011	**RO**	RK	*RK*	SU	975602	975603

CLASS 930/0 Sandite & De-icing Trailers

DT. Non gangwayed. Converted from Class 416/2.
Dimensions: 20.44 x 2.82 x 3.86 m. **Bogies**: Mark 3D.
Maximum Speed: 75 mph. **Doors**: Manually operated slam.
Couplings: Buckeye with additional Tightlock at non driving ends only.
Multiple Working: With Classes 317 and 319.

975578/579. DT. Dia. EZ526. Lot No. 30117 Eastleigh 1954. 32.5 t.

930 078	**RO**	RK	*RK*	HE	977578
930 079	**N**	RK	*RK*	SU	977579

CLASS 930/0 3-Car Route Learning Unit

DM–TB–DM. Gangwayed within unit. Converted from Class 411/4.
Supply System: 750 V d.c. third rail.
Traction Motors: Two English Electric 507 of 185 kW each per power car.
Dimensions: 20.34 x 2.82 x 3.83 m.
Maximum Speed: 90 mph. **Doors**: Manually operated slam.
Couplings: Buckeye. **Bogies**: Mk. 4/Commonwealth.
Multiple Working: SR type.

977861. DM. Dia. EZ536. Lot No. 30111 Eastleigh 1956. 44.2 t.
977862. TB. Dia. EZ542. Lot No. 30110 Eastleigh 1956. 36.2 t.
977863. DM. Dia. EZ536. Lot No. 30108 Eastleigh 1956. 43.5 t.

930 082	**CX**	SC	*OT*	SU	977861	977862	977863

CLASS 930/1 2-Car Tractor Unit

DMB–DMB. Gangwayed within unit.
Supply System: 750 V d.c. third rail.
Traction Motors: Two English Electric 507 of 185 kW each per power car.
Dimensions: 20.42 x 2.82 x 3.86 m.
Maximum Speed: 90 mph. **Doors**: Manually operated slam.
Couplings: Buckeye. **Multiple Working**: SR type.
Bogies: Mk. 4 or Mk. 3B/Commonwealth.

977207. DMB. Dia. EZ522. Lot No. 30388 Eastleigh 1958. 40.5 t.
977609. DMB. Dia. EZ522. Lot No. 30617 Eastleigh 1961. 40.5 t.

930 101	**N**	RK		AF(S)	977207	977609

CLASS 930/1 — 2-Car Sandite Unit

DMB–DMB. Gangwayed within unit. Converted from Class 416.
Supply System: 750 V d.c. third rail.
Traction Motors: Two English Electric 507 of 185 kW each per power car.
Dimensions: 19.23 x 2.74 x 3.99 m. **Doors**: Manually operated slam.
Maximum Speed: 75 mph. **Bogies**: Central 40 inch.
Couplings: Buckeye. **Multiple Working**: SR type.

977533. DMB. Dia. EZ512. Lot No. 4016 Eastleigh 1954–55. 40.5 t.
977534. DMB. Dia. EZ512. Lot No. 4099 Eastleigh 1955–56. 40.5 t.

930 102	**RO**	RK	*RK*	FR	977533	977534

CLASS 930/2 — 2-Car Sandite & De-icing Unit

DMB–DMB. Gangwayed within unit. Converted from Class 416/2.
Supply System: 750 V d.c. third rail.
Traction Motors: Two English Electric 507 of 185 kW each per power car.
Dimensions: 20.44 x 2.82 x 3.86 m. **Doors**: Manually operated slam.
Maximum Speed: 75 mph. **Bogies**: Mk. 3B.
Couplings: Buckeye. **Multiple Working**: SR type.

977566/567. DMB. Dia. EZ525. Lot No. 30116 Eastleigh 1954–55. 40.5 t.
977804/864. DMB. Dia. EZ522. Lot No. 30119 Eastleigh 1954. 40.5 t.
977805/865/871. DMB. Dia. EZ522. Lot No. 30167 Eastleigh 1955. 40.5 t.
977872/924/925. DMB. Dia. EZ522. Lot. No. 30314. Eastleigh 1956–58. 40.5 t.
977874/875. DMB. Dia. EZ522. Lot No. 30114 Eastleigh 1954. 40.5 t.

930 201	**RO**	RK	*RK*	FR	977566	977567
930 202	**RK**	RK	*RK*	FR	977804	977805
930 203	**RO**	RK	*RK*	RM	977864	977865
930 204	**N**	RK	*RK*	RM	977874	977875
930 205	**RO**	RK	*RK*	RM	977871	977872
930 206	**RO**	RK	*RK*	WD	977924	977925

CLASS 931 — 2-Car Route Learning Unit

DT–DMB or DMB–DT. Gangwayed within unit. Converted from Class 416/2.
Supply System: 750 V d.c. third rail.
Traction Motors: Two English Electric 507 of 185 kW each.
Dimensions: 20.44 x 2.82 x 3.86 m. **Doors**: Manually operated slam.
Maximum Speed: 75 mph. **Bogies**: Mk. 3B.
Couplings: Buckeye. **Multiple Working**: SR type.

977856. DT. Dia. EZ541. Lot No. 30168 Eastleigh 1955. 30.5 t.
977857. DMB. Dia. EZ522. Lot No. 30167 Eastleigh 1955. 40.5 t.
977917. DMB. Dia. EZ541. Lot No. 30119 Eastleigh 1954. 40.5 t.
977918. DT. Dia. EZ541. Lot No. 30120 Eastleigh 1954. 30.5 t.

931 001	**N**	SE	*OT*	RM	977856	977857
931 002	**N**	SE	*OT*	RM	977917	977918

CLASS 931 2-Car Tractor Unit

DMB–DMB. Gangwayed within unit. Converted from Class 416/2.
Supply System: 750 V d.c. third rail.
Traction Motors: Two English Electric 507 of 185 kW each per power car.
Dimensions: 20.44 x 2.82 x 3.86 m. **Doors**: Manually operated slam.
Maximum Speed: 75 mph. **Bogies**: Mk. 3B.
Couplings: Buckeye. **Multiple Working**: SR type.

977559/560. DMB. Dia. EZ525. Lot No. 30116 Eastleigh 1954-55. 40.5 t.

931 062	**N**	SE	*OT*	RM	977559	977560	The Sprinkler

CLASS 931 Tractor Unit

DM. Non gangwayed. Previously Class 419.
Supply System: 750 V d.c. third rail or battery power.
Traction Motors: Four English Electric 507 of 185 kW each per power car.
Dimensions: 19.64 x 2.82 x 3.86 m. **Doors**: Manually operated slam.
Maximum Speed: 90 mph. **Bogies**: Mk. 3B.
Couplings: Buckeye. **Multiple Working**: SR type.

68002. DM. Dia. EX560. Lot No. 30458 Eastleigh 1959. 45.5 t.
68003–68010. DM. Dia. EX560. Lot No. 30623 Eastleigh 1960–61. 45.5 t.

9003	**B**	P	BM(S)	68003
9007	**J**	P	BM(S)	68007
9009	**J**	P	BM(S)	68009
931 090	**J**	P	BM(S)	68010
931 092	**N**	P	BM(S)	68002
931 094	**N**	P	BM(S)	68004
931 095	**J**	P	BM(S)	68005
931 098	**N**	P	BM(S)	68008

CLASS 932 3-or 4-Car Test Units

DM–TB–T–DM. Gangwayed throughout. Converted from Class 411. Test units for manufacturers' traction packages.
Supply System: 750 V d.c. third rail (* or 15 kV a.c. 16.67 Hz overhead; † or 25 kV a.c. 50 Hz overhead).
Traction Motors: Adtranz. († Two English Electric 507 of 185 kW each (61948); Alstom (61949).
Dimensions: 20.34 x 2.82 x 3.83 m.
Maximum Speed: **Doors**: Manually operated slam.
Couplings: Buckeye. **Multiple Working**: SR type.
Bogies: Mk. 4 /Commonwealth.
Non Standard Liveries:

- 932 545 is Adtranz blue with a white stripe.
- 932 620 has one side of each car painted in GEC-Alsthom white and orange livery and the other side of each car painted in livery **P**.)

Note: 932 545 is currently based at the Adtranz works at Västerås, Sweden.

61358/359. DM. Dia. EZ???. Lot No. 30454 Eastleigh 1958–59.
61948/949. DM. Dia. EZ???. Lot No. 30708 Eastleigh 1963.
70330. TB. Dia. EZ???. Lot No. 30456 Eastleigh 1958–59.
70653. TB. Dia. EZ???. Lot No. 30709 Eastleigh 1963.

932 545	*	**0**	P	*AD*	Sweden	61359	70330	61358	
932 620	†	**0**	P	*AM*	IL	61948	70653	70660	61949

CLASS 936/0 — 2-Car Sandite Unit

DM–DTB. Gangwayed throughout. Converted from Class 501.
Supply System: 750 V d.c. third rail.
Traction Motors: Four GEC of 137 kW each.
Dimensions: 18.47 x 2.90 x 3.86 m.
Doors: Manually operated slam.
Maximum Speed: 70 mph.
Bogies: BR2.
Couplings: Buckeye.
Multiple Working: SR type.

977349. DM. Dia. EZ504. Lot No. 30326 Eastleigh 1957–58. 48.0 t.
977350. DTB. Dia. EZ506. Lot No. 30328 Eastleigh 1957–58. 30.5 t.

936 003	**MD**	RK	*RK*	BD	977349	977350

CLASS 936/1 — 3-Car Sandite Unit

DT–MB–DT. Non gangwayed. Converted from Class 311.
Supply System: 25 kV a.c. 50 Hz overhead.
Traction Motors: Four AEI of 165 kW each.
Dimensions: 20.18 x 2.82 x 3.86 m.
Maximum Speed: 75 mph.
Doors: Power operated sliding.
Couplings: Buckeye.
Bogies: Gresley ED3/ET3.
Multiple Working: Classes 303–312.

977844/847. DT. Dia. EZ543. Lot No. 30767 Cravens 1967. 34.4 t.
977845/848. MB. Dia. EZ544. Lot No. 30768 Cravens 1967. 56.4 t.
977846/849. DT. Dia. EZ543. Lot No. 30769 Cravens 1967. 38.4 t.

936 103	**RO**	RK	*RK*	GW	977844	977845	977846
936 104	**RO**	RK	*RK*	GW	977847	977848	977849

CLASS 937 — 3-Car Sandite Unit

BDT–MB–DT. Gangwayed within unit. Converted from Class 308.
Supply System: 25 kV a.c. 50 Hz overhead.

For details see Class 308.

977876/926. BDT. Dia. EZ545. Lot No. 30656 York 1961. 36.3 t.
977877/927. MB. Dia. EZ546. Lot No. 30657 York 1961. 55.0 t.
977878/928. DT. Dia. EZ547. Lot No. 30659 York 1961. 33.0 t.

937 990	**N**	RK	*RK*	EM	977876	977877	977878
937 991	**N**	RK	*RK*	IL	977926	977927	977928

UNCLASSIFIED Generator Coach

QXA. Gangwayed throughout. Converted from Class 438. Operates with 999550.
Dimensions: 20.18 x 2.82 x 3.81 m.
Maximum Speed: 90 mph. **Doors:** Manually operated slam.
Couplings: Buckeye. **Bogies:** B5 (SR).

977335. QXA. Dia. QX174. Lot No. 30764 York 1966–67. 32.0 t.

– **S0** SO *SO* ZA 977335

UNCLASSIFIED LUL Track Recording Car

T. LUL Track Recording Car. Converted from 1973 tube stock.
Dimensions:
Maximum Speed: 70 mph. **Doors:** Power operated sliding.
Couplings: Buckeye. **Bogies:** LT design.
Non Standard Livery: White with blue lower body stripe and red doors.
Note: Also carries LUL number TRC666.

999666. T. Dia. EZ548. Met-Camm. 1974. Converted BREL Derby 1987. 23.8 t.

– **0** LU *OT* WR 999666

5. VEHICLES AWAITING DISPOSAL

25 kV a.c. 50 Hz OVERHEAD EMUS

Complete Units:

302 201	**N**	H	PY(S)	75085	61060	70060	75033
302 204	**N**	H	PY(S)	75088	61063	70063	75036
302 213	**N**	H	PY(S)	75097	61072	70072	75060
302 216	**N**	H	PY(S)	75100	61075	70075	75063
302 218	**N**	H	PY(S)	75191	61077	70077	75065
302 221	**N**	H	PY(S)	75194	61080	70080	75068
302 224	**N**	H	PY(S)	75197	61083	70083	75071
302 225	**N**	H	PY(S)	75198	61084	70084	75072
302 226	**N**	H	PY(S)	75199	61085	70085	75073
302 227	**N**	H	PY(S)	75325	61193	70193	75250
302 228	**N**	H	PY(S)	75201	61087	70087	75075
302 230	**N**	H	PY(S)	75205	61091	70091	75079

Spare Car:

Cl. 307	**BG**	E	KN(S)	75023

750 V d.c. THIRD RAIL EMUS

Complete Units:

4308	**N**	H	PY(S)	61275	75395		
5001	**G**	H	KN(S)	14001	15207	15101	14002
5176	**B**	H	KN(S)	14352	15396	15354	14351
6213	**BG**	H	PY(S)	65327	77512		
6308	**N**	H	PY(S)	14564	16108		
6309	**N**	H	PY(S)	14562	16106		
6402	**N**	H	PY(S)	65362	77547		
7001	**N**	P	ZG(S)	67300	67301		

LIVERY CODES

* denotes an obsolete livery style no longer used for repaints.

Code	*Description*
AL	Advertising livery *(See class heading)*.
B*	BR *(Blue)*.
BG*	BR *(Blue and grey lined out in white)*.
CC	BR/Strathclyde PTE *(Carmine & cream lined out in black and straw)*.
CO	Centro *(Grey/green with light blue, white & yellow stripes)*.
CS	Connex *(Blue with yellow lower body and blue solebar)*.
CX	Connex *(White with yellow lower body and blue solebar)*.
EU	Eurostar. *(White with dark blue & yellow stripes)*.
G*	BR *(Plain or two-tone green)*.
GE	First Great Eastern *(Grey, green, blue and white)*.
GM	Greater Manchester PTE *(Light grey/dark grey with red and white stripes)*.
GX	Gatwick Express *(Dark grey/white/burgundy/white)*.
HX	Heathrow Express *(Silver with black window surrounds)*.
J*	London & South East sector *(Two tone brown with orange stripe)*.
LS	LTS Rail *(Grey/white/green/white/blue/white)*.
MD	Mersey Travel departmental *(Yellow/black)*.
MT	Mersey Travel *(Yellow/white with grey and black stripes)*.
N*	BR Network South East *(Grey/white/red/white/blue/white)*.
NT*	Network South East *(Grey/red/white/blue/white)*.
NW	North West Trains *(Blue with gold cant rail stripe and star)*.
0	Non standard liveries *(see notes in class headings for details)*.
PM*	Provincial Midline *(Dark blue/grey with dark blue & grey stripes)*.
RK	Railtrack *(Green and blue)*.
RM	Royal Mail *(Red with yellow stripes)*.
RN*	North West Regional Railways *(Dark blue/grey with green & white stripes)*.
RO*	Railtrack *(Orange with white and grey stripes)*.
RR*	Regional Railways *(Dark Blue/Grey with light blue & white stripes, three narrow dark blue stripes at cab ends)*.
S*	Strathclyde PTE *(Orange/black lined out in white)*.
SL	Silverlink *(Indigo blue with white stripe, green lower body & yellow doors)*.
ST	Stagecoach South West Trains *(White/orange/white/red/white/blue/white)*.
SW	South West Trains *(White/black/orange/red/blue, with red doors)*.
TR	Thameslink Rail *(Dark blue with a broad orange stripe and two narrower white bodyside stripes plus white cantrail stripe)*.
U	Undercoat *(White or grey undercoat)*.
WN	West Anglia Great Northern Railway *(White with blue, grey and orange stripes)*.
WY*	West Yorkshire PTE *(Red/cream with thin yellow stripe)*.

OWNER CODES

Code	*Owner*
A	Angel Train Contracts Ltd.
CM	Cambrian Trains Ltd.
E	English Welsh & Scottish Railway Ltd.
EU	Eurostar (UK) Ltd.
H	HSBC Rail (UK) Ltd.
HX	British Airports Authority PLC.
LU	London Underground Ltd.
P	Porterbrook Leasing Company Ltd.
RK	Railtrack PLC.
RM	Royal Mail.
SB	Belgian National Railways.
SC	Connex South Central.
SE	Connex South Eastern.
SF	French National Railways.
SO	Serco Railtest Ltd.

OPERATION CODES

Code	*Operator*
AD	Adtranz.
AM	Alstom.
CT	Central Trains.
E	English Welsh & Scottish Railway.
EU	Eurostar (UK).
GE	First Great Eastern.
GX	Gatwick Express.
HX	Heathrow Express.
IL	Island Line.
LS	LTS Rail.
ME	Merseyrail Electrics.
NS	Northern Spirit.
NW	First North Western.
OT	Non revenue earning (e.g. Test Trains, Research, Route Learning).
RK	Railtrack.
SC	Connex South Central.
SE	Connex South Eastern.
SL	Silverlink.
SO	Serco Railtest.
SR	ScotRail.
SW	South West Trains.
TR	Thameslink Rail.
WN	West Anglia Great Northern.

DEPOT TYPE ABBREVIATIONS

CARMD	Carriage Maintenance Depot
SD	Servicing Depot
T&RSMD	Traction & Rolling Stock depot

DEPOT & LOCATION CODES

* denotes unofficial code.

Code	*Location*	*Operator*
AF	Chart Leacon (Ashford, Kent) T&RSMD	Adtranz
BD	Birkenhead North T&RSMD	Merseyrail Electrics
BI	Brighton T&RSMD	Connex South Central
BM	Bournemouth T&RSMD	South West Trains
BY	Bletchley T&RSMD	Silverlink
CB*	Crewe Brook Sidings	Storage location only
CE	Crewe International Electric T&RSMD	EWS
CJ	Clapham Junction (London) SD	South West Trains
EM	East Ham T&RSMD	LTS Rail
FF	Forest (Brussels) T&RSMD	SNCB/NMBS
FR	Fratton T&RSMD	South West Trains
GI	Gillingham (Kent) T&RSMD	Connex South Eastern
GW	Shields Road (Glasgow) T&RSMD	ScotRail/Alstom
HE	Hornsey T&RSMD	West Anglia Great Northern
IL	Ilford T&RSMD	First Great Eastern
KK	Kirkdale SD	Merseyrail Electrics
KN*	MOD Kineton	Storage location only
LG	Longsight Electric (Manchester) T&RSMD	Cross Fleet
LY	Le Landy (Paris) T&RSMD	SNCF
NB*	New Brighton EMU Sidings	Storage location only
NL	Neville Hill DMU/EMU (Leeds) T&RSMD	Northern Spirit
NP	North Pole International (London) T&RSMD	Eurostar (UK)
OH	Old Oak Common Electric (London) T&RSMD	Siemens
PY*	MOD Pig's Bay (Shoeburyness)	Storage location only
RM	Ramsgate T&RSMD	Connex South Eastern
RY	Ryde (Isle of Wight) T&RSMD	Island Line
SG	Slade Green T&RSMD	Connex South Eastern
SI	Soho (Birmingham) T&RSMD	Maintrain
SL	Stewarts Lane (London) T&RSMD	Gatwick Express
SU	Selhurst (Croydon) T&RSMD	Connex South Central
WD	Wimbledon T&RSMD	South West Trains
WK*	West Kirkby	Storage location only
WR*	West Ruislip T&RSMD	London Underground
ZA	Railway Technical Centre, Derby	Serco Railtest/Fragonset Railways
ZD	Litchurch Lane, Derby	Adtranz
ZG	Eastleigh	Alstom
ZH	Springburn	Railcare
ZN	Wolverton	Railcare

PLATFORM 5 PUBLISHING LTD MAIL ORDER

Modern British Railways Titles	Price
BR Pocket Book 1: Locomotives 42nd edition	£2.75
BR Pocket Book 2: Coaching Stock 24th edition	£2.75
BR Pocket Book 3: DMUs & Light Rail 13th edition	£2.75
BR Pocket Book 4: Electric Multiple Units 13th edition	£2.75
Diesel & Electric Loco Register 3rd edition	£7.95
Preserved Coaching Stock Part 1 1st edition	£7.95
Preserved Coaching Stock Part 2 1st edition	£8.95
Valley Lines - The People's Railway	£9.95
British Rail Depot Directory 3rd edition	£6.95
BR Wagon Fleet-Air Braked Freight Stock	£7.95
BR Wagon Fleet-B Prefix Freight Stock	£6.95
Engineers Series Wagon Fleet	£6.95
Departmental Coaching Stock 5th edition	£6.95
On-Track Plant on British Railways	£7.95
Signalling Atlas & Directory	£9.95
Miles & Chains Scottish Region	£1.95
Miles & Chains Southern Region	£1.95
British Rail Mark 2 Coaches	£29.99

BR Locomotives & Coaching Stock Back Numbers

BR Locomotives & Coaching Stock 1999	£10.75
BR Locomotives & Coaching Stock 1998	£10.50
BR Locomotives & Coaching Stock 1997	£9.95
BR Locomotives & Coaching Stock 1996	£8.95
BR Locomotives & Coaching Stock 1995	£8.50
BR Locomotives & Coaching Stock 1994	£7.50
BR Locomotives & Coaching Stock 1993	£7.25
BR Locomotives & Coaching Stock 1992	£7.00
BR Locomotives & Coaching Stock 1991	£6.60
BR Locomotives & Coaching Stock 1990	£5.95
BR Locomotives & Coaching Stock 1989	£4.95
BR Locomotives & Coaching Stock 1988	£3.95
BR Locomotives & Coaching Stock 1987	£3.30
BR Locomotives & Coaching Stock 1986	£3.30

Overseas Railways

Benelux Railways 3rd edition £10.50
Austrian Railways 3rd edition £10.50
French Railways 3rd edition £14.50
Swiss Railways 2nd edition £13.50
Italian Railways 1st edition £13.50
Irish Railways 1st edition £9.95
High Speed in Europe £9.95
High Speed in Japan £16.95
Railways of Boston £8.25
The Railways of Greece £8.10
The Railways of Corsica £5.10
The Railways of Tunisia £8.50
The Railways & Tramways of Hong Kong £8.95
The Railways of Chile: Volume 1 .... £8.95
Railways Of Sardinia £8.25
Hungarian Motive Power £16.95
The Light Track From Arras £8.95
Narrow Gauge Railways of Portugal £24.95
Forestry Railways in Hungary £9.95
Belfast & County Down Railway £12.99
Dublin & South Eastern Railway £19.99
Vivarais Narrow Gauge £12.95
Réseau Breton £10.95
Swiss Mountain Railways Volume 1 .... £12.95
Electric & Diesel Locomotives of South Africa £10.95
Steam & Rail in Slovakia £9.95
Steam & Rail in Germany £9.95
Broader Than Broad £5.95
Soviet Locomotive Types £12.95
A Guide To Indian Railways £8.95
Rails in the Austrian Tirol £8.95
Railways around Lake Luzern £9.95
Narrow Gauge Railways of Spain Volume 1 .... £10.95
Narrow Gauge Railways of Spain Volume 2 .... £10.95
Mountain Rack Railways of Switzerland £9.95
TGV Handbook 2nd edition £10.95
Narrow Gauge Rails to Esquel £9.95
Amsterdam Utrecht As It Was £24.95

Light Rail Transit, Trams & Metro Systems

Light Rail Review 3 .... £7.50
Light Rail Review 4 .... £7.50
Light Rail Review 5 .... £7.50
Light Rail Review 6 .... £7.50
Light Rail Review 7 .... £8.95
Light Rail Review 8 .... £9.50
Blackpool & Fleetwood 100 Years by Tram £19.95
Manx Electric £8.95
Freiburg: Classic Tramway to Light Rail £17.70
The Tramways of Portugal £9.05
Paris Metro Handbook 3rd edition £9.95
London Underground Rolling Stock £9.95
Underground Official Handbook £7.95
Docklands Official Handbook £7.95
The Twopenny Tube £5.95
The Piccadilly Line £5.95
The Northern Line £5.95
Circles Under the Clyde £15.95
Underground Architecture £25.00
Mr. Beck's Underground Map £12.95
Pleasure Trips by Underground £14.95
World Metro Systems £10.95
The Berlin S-Bahn £7.50
The Berlin U-Bahn £7.50
Light Rail in Europe £9.95
London Tramways £19.95

Maps, Atlases & Track Diagrams

Railway Track Diagrams:1 Scotland & IOM £6.50
Railway Track Diagrams:2 England East £7.95
British Rail Track Diagrams:4 London Midland (1990 Reprint) £6.95
Railway Track Diagrams:6 Ireland £5.50
Railway Track Diagrams:7 New South Wales £6.50
London Transport Railway Track Map £1.75
China Railway Atlas £10.00
MR System Maps: Settle–Carlisle £4.95
MR System Maps: Leeds–Leicester £9.95
MR System Maps: Leicester–London £8.95

MR System Maps: Birmingham–Bristol £7.95
MR System Maps: Scotland & Index £7.95
MR System Maps: The Gradient Profiles £7.50
Johnson's Atlas & Gazetteer £19.99
European Railway Atlas: Germany Denmark Austria Swiss £13.99
World Railway Gauge Map £5.95
Europe Railway Electrification Map £6.95
Slovenian Railways Tourist Map £7.95

Nostalgia

6203 Princess Margaret Rose £19.95
Rails along the Sea Wall £4.95
Steam Days on BR 1:Midland Line in Sheffield £4.95
The Rolling Rivers £6.95
British Baltic Tanks £6.95
British Railway Locomotives 1948–60 .... £18.95
The Hunslet Engine Works £25.00
Bradshaw's Railway Manual 1923 .... £12.50
Bradshaw's Guide 1863 .... £9.95
London Tilbury & Southend Railway Part 1 .... £9.95
London Tilbury & Southend Railway Part 2 .... £9.95
The Thames Haven Railway £10.95
Beyer Peacock Locomotive Builders £32.50
Suburban Railways of Tyneside £9.95
Scenes from the Past 29: Woodhead Part 2 .... £14.95
London's Branch Railways £19.99
Railways around London: Then & Now £34.99
Davies & Metcalfe Ltd £30.00
Sussex Steam £7.95
London Local Railways £25.00
The Rise of the Midland Railway £11.95
South for Sunshine £7.95

Rambling Guides

Rambles by Rail 2:Liskeard–Looe £1.95
Rambles by Rail 4:The New Forest £1.95

PVC Book Covers (Available in Blue, Red, Green or Grey)

A6 Pocket Book Plastic Cover £1.00
A5 Size Plastic Cover £1.50

HOW TO ORDER

Telephone your order and credit card details to our 24-hour sales hotline:

0114 255 8000 (UK), +44 114 255 8000 (from overseas).
An answerphone is attached for calls made out of office hours.

Or, fax to:

0114 255 2471 (UK), +44 114 255 2471 (from overseas).

We accept Credit/Debit Card payments by Visa, MasterCard, Eurocard, Delta & Switch. Please state type of card, card number, issue no./date (for Switch Cards only), expiry date, and full name & address of card holder.

!!! Please note - the minimum credit card order accepted is £3.00!!!

Or, send your Credit/Debit Card details, Sterling Cheque , Eurocheque, Money Order or British Postal Order payable to '**Platform 5 Publishing Ltd.**' to:

Mail Order Department (PB)
Platform 5 Publishing Ltd.
3 Wyvern House, Sark Road
SHEFFIELD, S2 4HG, ENGLAND

Please add postage & packing: 10% UK; 20% Europe; 30% Rest of World, 40p minimum. If p&p works out at less than 40p, then please add on 40p, this is the minimum p&p accepted.

NOTE: Overseas postal prices are for despatch by Air Freight. Transit time: Europe 2 weeks; Rest of World 4 weeks.

For a full list of titles available from Platform 5 Mail Order, please send a SAE to the above address.

Vouchers: When ordering publications in conjunction with a **Today's Railways** subscription offer please add on post and packing **before** deducting the voucher. Vouchers may **not** be combined.

Details correct as at 23rd November 1999. Prices are not guaranteed and we reserve the right to alter details without further notification. Please allow 28 days for delivery in the UK.